THE ILLUSTRATED
THEORY OF EVERYTHING

The Origin and Fate of the Universe

THE ILLUSTRATED

THEORY OF EVERYTHING

The Origin and Fate of the Universe

STEPHEN W. HAWKING

PHOENIX
BOOKS

Copyright © 2003 by New Millennium Press
Text first published under the title The Cambridge Lectures: Life Works
Copyright © 1996 by Dove Audio, Inc.

Phoenix Books, Inc.
9465 Wilshire Boulevard, Suite 840
Beverly Hills, CA 90212

Library of Congress Cataloging-in-Publication Data available upon request.

ISBN: 978-1-59777-611-0

Cover design: Sonia Fiore
Book design: Rob Fiore

Jacket Photo credit: Megastar-Birth Cluster is Biggest, Brightest and Hottest Ever Seen
NASA Hubble Space Telescope Collection
ESA [http://spacetelesco…], Hubble European Space Agency
[http://spacetelesco…] Information Centre (M. Kornmesser
and L.L. Christensen), and NASA [http://www.nasa.gov/]

Printed in the United States of America

Phoenix Books, Inc.
9465 Wilshire Boulevard, Suite 840
Beverly Hills, CA 90212

10 9 8 7 6 5 4 3 2 1

THEORY OF EVERYTHING:
A FOREWORD

By Marcelo Gleiser, PhD
Appleton Professor of Natural Philosophy
Professor of Physics and Astronomy, Dartmouth College

The most remarkable property of the universe is that it has spawned creatures able to ask questions. I know that some may consider this statement to be anthropocentric, elevating human curiosity to an undeserved level of centrality in the (very) big scheme of things. "There may be other intelligent civilizations out there, asking questions about their origins," a reader may protest. True, there may be other curious entities somewhere in our galaxy or in other galaxies across the cosmos. My opening statement includes them too. However, UFO-sighting claims to the contrary, we haven't heard from them yet and chances are we won't for a very long while.

The first thing we learn about the universe is how vast it is. As an illustration, consider our nearest stellar neighbor, Alpha Centauri, at about 4.5 light years from us. That is, it takes 4.5 years for light from the star to reach us, traveling at the mind-boggling speed of three hundred thousand kilometers per second. Our fastest spaceship would take over one hundred thousand years to get there. And this is our nearest neighbor! Even though other intelligent life forms may exist, and contact could one day happen, it's safe to assume that while we may not be alone wondering about the mysteries of creation, we are alone seeking for answers. For all we know, we are how the universe thinks about itself. And we have been quite busy.

The history of civilization can be told in many ways. One version could focus on trade; another could focus on technology, or how increasingly sophisticated tools allowed for energy to be harnessed in more efficient ways; yet another could focus on our ideas about Nature and how they shifted in time, along with global trade and energy-extracting tools. The universe we live in today is very different from the universe of a fifteenth century person. Columbus sailed west convinced he lived in an Earth-centered cosmos, with crystalline spheres carrying the planets about in their circular orbits. Of course, the universe itself hasn't changed much in the past five hundred years. But the way we think about it has. In *The Theory of Everything*, Stephen Hawking offers a masterful guide to the cosmos *circa* year 2000. He is well aware that the guide will be edited as years go by, that the only constant as we explore the universe is that our views about it will change. It may hurt our pride to admit this, but the truth is that in five hundred years our present views will also be seen as primitive. Important, to be sure, as science always builds up on past knowledge to advance. But primitive nevertheless.

One thing, though, has not and will not change: our need to understand our origins, to address the mystery of creation. Ever since the first hominids gathered in groups, they must have wondered about the workings of Nature. Our oldest artifacts and cave paintings depict symbols of fertility and of animal diversity and worship. Our distant ancestors made Nature sacred, explaining both its regular cycles and its unpredictability in terms of the will of gods. Rituals established channels of communication, whereby gods could be appeased or taunted into action. In this pre-scientific, pre-philosophical era, to understand Nature was to worship it. Mythic narratives told of how the gods interfered with the world and with the affairs of men, constituting the roots of what, much later, was to become science. These stories tried to make sense of the unknown through familiar images everyone could relate to. So, one god

carried the Sun about the sky in a fiery chariot, another lit up
the stars at night, while another controlled the rains. The stories
changed from place to place, reflecting the local environment
and climate. But their essence, their main themes, remained the
same across the globe: how the world came to be, how people
and animals came to be, why we die, will the world come to an
end. These creation myths were the first cosmologies, the first
explanations of the workings of the cosmos. Even if our modern
scientific explanations are radically different from those of our
ancestors, they belong to the same ancient tradition, reflecting
the same need to understand our origins.

With the advent of western philosophy in ancient Greece,
the focus changed from the mythic to the rational. Around 650
B.C.E., Thales, considered by none other than Aristotle to be
the first philosopher, wanted to know what the world was made
of. He believed that all matter could be traced back to a single
substance, the one stuff common to everything. His answer?
Water. Do not judge the accuracy of the answer with your
twenty-first century values. The important point here is that
Thales believed in a unified description of Nature. In suggesting
that a single substance described the material world, he was
effectively proposing a theory of everything, the *first* theory
of everything. The roots of our modern search for a theory of
everything reach far back in time.

Thales and his followers believed that Nature's essence was
change. Theirs was a philosophy of becoming, of transformation.
Everything emerged and returned to the fundamental substance,
in an eternal dance of creation and destruction. As usual in
philosophy, not everyone shared these views. Parmenides, in
Italy, devised a contrary opinion, that what is essential cannot
change. If you were interested in the true nature of things, focus
on what is eternal, unchangeable, and not on what is ephemeral.
His was a philosophy of being. This tension between being and
becoming lies at the heart of science. We search for unchangeable

laws of Nature, blueprints to all the changes we observe in the form of natural phenomena. As Hawking writes, "our aim is to formulate a set of laws that will enable us to predict events up to the limit set by the uncertainty principle." So, science is built upon the complementarity of being and becoming: the (presumably) unchanging laws of Nature describe all variety of natural phenomena.

The key ingredient in this formulation is mathematics. Physical laws are expressed in mathematical language. Again, this tradition has its roots in ancient Greece. While Thales and his followers argued with Parmenides about being and becoming, Pythagoras was founding a mystical-philosophical society in southern Italy with the goal of expressing all of Nature in terms of numbers. Geometry, they believed, was the essence of all things. Forms and shapes describe all there is, and they, in turn, are described by numerical relationships. Uncovering these relations was to understand the mind of God: God is a geometer, and the goal of science is to know His mind. This image is used quite openly nowadays in scientific circles, albeit mostly as a metaphor. Hawking himself uses it in the closing lines of this book. To ask deep questions about Nature is, ultimately, to want to know the mind of God.

As science evolved, the search for a theory of everything persisted. During the Renaissance, the German astronomer Johannes Kepler—a true Pythagorean—searched for the geometrical structure of the cosmos, a sort of map of creation based on simple mathematical rules. Although he failed, he did find the three laws of planetary motion along the way, including the famous one stating that planetary orbits are elliptical and not circular. As often happens in science, discoveries are made in the pursuit of an elusive (and sometimes nonexistent) goal. The great Isaac Newton, who brought Galileo's terrestrial physics and Kepler's celestial physics into the umbrella of a single theory of gravity, also searched for a theory of everything. Newton viewed the mathematical nature of the universe as a manifestation of the mathematical mind of God. To understand Nature was, quite literally, a religious mission, to decipher God's plans for the world. His three laws of motion and his law of universal gravity were a major step in proving the existence of unchangeable, mathematical laws that work throughout the cosmos. Newton demonstrated that human reason could grapple with the mystery of creation. Suddenly, Nature became an open book and the natural philosopher (the old name for scientists) its interpreter.

During the eighteenth and nineteenth centuries, scientists forged an even deeper belief in the power of mathematics to describe natural phenomena. From ocean tides to the orbits of planets and comets, from the motion of fluids to the property of gases and the industrial use of steam, from the deep relationship between electricity and magnetism to the invention of electric motors, Newtonian science and its spinoffs changed the world. And yet, the belief that behind the staggering variety of natural phenomena there exists a simple underlying mathematical structure remained as alive as ever. Michael Faraday, one of the greatest physicists of all time, died convinced that gravity and electromagnetism were aspects of the same fundamental force. He searched for signs of this unification in his laboratory for years, admitting to having failed in practice but never wavering in his beliefs.

Physics underwent a deep transformation during the first three decades of the twentieth century. The successes of the past remained valid, of course, but a new physics emerged out of the pressing need to explain the bizarre world of the atom. The old views had to be abandoned, as certainty and determinism gave way to uncertainty and probabilities. Quantum mechanics, as the new physics of the very small was called, was a major departure from Newtonian physics. Gone were the familiar phenomena of rocks falling and currents flowing in wires. Electrons could be both particles and waves, could be here and there, and in a strict sense only existed if someone measured them: the observer became deeply linked with the observed. Behind it all stood Heisenberg's uncertainty principle, which stated that it is impossible to measure both the position and the velocity of a particle with arbitrarily high accuracy. The principle imposed a limit on how precise our measurements can be and on how much information we can extract from an experiment. Hence Hawking's declaration, previously quoted, accepting "the limit set by the uncertainty principle," as the limit of our understanding of natural phenomena.

Lest the reader condemns quantum mechanics as an unfortunate development in the history of science, let me assure you that in fact it's quite the opposite. Most of modern society depends in one way or another on our understanding of the quantum properties of matter. All modern digital technology, lasers, and the nuclear and electronics revolution of the 1950s are by-products of the physics of the very small. This physics, as Hawking describes in *The Theory of Everything*, would also launch a new era of cosmology, where the most mysterious of questions, the origin of the universe as a whole, left the realm of myth and entered the realm of science. Creation—the last scientific frontier.

In 1929, the American astronomer Edwin Hubble discovered that the universe is expanding: distant galaxies are moving away from one another. Although our first impulse is to picture them as shrapnel flying away from an explosion (the big bang), the correct way to think about the expanding universe is to imagine that space itself is being stretched as if it were a rubber balloon. The galaxies are just riding along. This plasticity of space was described in Einstein's general theory of relativity, proposed in 1915: matter causes space to curve around it. According to the theory, the effects of gravitational attraction can be understood as accelerated motion in curved geometries. Going backwards in time, we can envision a point in the past when all the galaxies were atop one another, squeezed into a tiny volume. At this point, known as the "singularity," the laws of physics break down. At least the laws of classical (non-quantum) physics, as Hawking and Oxford's Roger Penrose demonstrated in the sixties and seventies. This should not come as a surprise. After all, if the universe as a whole were of atomic dimensions in its infancy, quantum mechanics must have dictated what was happening. A proper scientific description of the big bang can only be obtained once we know how to combine quantum mechanics with Einstein's theory of gravity. Unfortunately, so far we haven't been able to do this. Fortunately, we have physicists like Stephen Hawking to pave the way. As only visionaries can, Hawking saw imprints of the true quantum gravity world lying among the confusion chaos of creation.

The first clue came from black holes. As the name says, black holes are expected to be black, that is, they are not supposed to emit light or any other form of radiation. According to Einstein's general relativity, a black hole would form when space bends around itself so much that nothing, not even light, can escape it. As everyone knows, the steeper the hill, the harder it is to climb. For someone walking uphill, there will be a degree of steepness beyond which walking up becomes impossible. The same happens near a black hole. Whatever is trapped within its horizon, a sort of spherical boundary around it, cannot escape to the outside world. Little black holes have small horizons and large black holes have large ones. If the Sun were to turn into a black hole (relax, it won't), its horizon radius would be a mere 3 km. What Hawking realized was that if you apply the basic laws of quantum mechanics to a black hole, it could not be perfectly black. Quantum uncertainty would play a role near the strange creatures, predicting that particles could be created in pairs out of random energy fluctuations. If one were to be sucked in, the other would have a chance of escaping far away. Black holes can "sweat," or better, evaporate. Hawking showed that the rate at which black holes emit particles is inversely proportional to their masses: the heavier the black hole, the less it sweats, kind of the opposite of humans. As Hawking remarks in his book, the effect is small and hard to observe. But it does predict that black holes have a finite lifetime, even though for most of the large ones this lifetime is longer than the age of the universe.

The second clue into the workings of quantum gravity came from Hawking's work on what became known as "quantum cosmology," where quantum physics is applied to the universe as a whole. Together with James Hartle, from the University of California in Santa Barbara, Hawking proposed that the initial state of the universe was one of timelessness. Nothing ever happened; the universe just was. "The universe would be neither created nor destroyed. It would just be," he wrote. Parmenides would be proud. Here is a modern scientific version of the universe of being. Ironically, Hawking put forward this idea during a conference in the Vatican, not yet knowing that it would have serious theological implications. For one thing, a universe that is neither created nor destroyed, that is, a universe that is uncreated, does not need a Creator. What would be the role of God if science could explain creation itself?

Of course, the universe does evolve in time. The universe of being, as Hawking explains, exists only in what mathematicians call "imaginary time." In this formulation, the universe of being remains as is, ignorant of change for all eternity. However, in real time, singularities do exist and call for the breakdown of the laws of physics. Which one is it? Could it be that imaginary time is the correct choice and that what we call time is just a tool we need to quantify change? In reality, it doesn't matter. The beauty of mathematics is that it allows us to explore relationships between ideas and the real world. As one goes from imaginary time to real time, the universe of being morphs into a universe of becoming. Time as we know it emerges at the transition from quantum to classical, and the cosmos begins to change without the complexities of an initial singularity.

A universe of being, where no change ever happens, is a self-contained entity that needs no creator, while a universe with time and thus with a singularity always has room for God. So, God or no God? There is no resolution for this enigma, at least for now. However, Hawking admits that science can only go so far in explaining everything. Even in a universe where time is imaginary, "[God] still had the freedom to choose the laws that the universe obeyed." Science, at least as it exists now, cannot breach this limitation. There is nothing new here. In fact, this is an old philosophical dilemma, known as the problem of the First Cause. If everything with a beginning has a cause, then what caused the universe? Although science got much closer to the actual moment of creation, it still can't move all the way into it. Religions get away with this by making use of supernatural arguments such as "God exists outside time and hence is uncreated and uncaused." It is quite hard to argue with statements like this while sticking to naturalist (as opposed to supernaturalist) reasoning. So, the limitation of science, at least from the point of view of understanding the origin of the universe, is that it can't cheat itself. In my opinion, it is best to admit that science is a narrative we create to describe Nature, and as such has no obligation or commitment to explain what is beyond its jurisdiction. Whatever theory we come up with, it will necessarily rely on a scientific framework that makes use of hypotheses, mathematics, and a handful of fundamental physical laws. Asking science to answer nonscientific questions forces it into a defensive position which is completely unwarranted. Instead, we should be in awe of all that science has accomplished in only four hundred years, thanks to a large extent to minds such as Hawking's.

All we can do is inch forward, searching for more clues toward the ultimate theory of everything. But why should such a theory exist? Or better, how do we know such a theory exists? And what does it mean to have a theory of everything? As I noted earlier, the search for the theory of everything is as old as science itself. Ever since Thales proposed his solution ("all is water"), and Faraday looked for it in his laboratory, we have been searching for the unified explanation for all that is. Religions found it in God; science tries to find it in the theory of everything. There is much optimism that such a theory exists, in spite of the current paucity of experimental data pointing towards it. Einstein, for one, spent decades looking for a unified description of gravity and electromagnetism. True, there are a few clues that the forces of Nature (gravity, electromagnetism, and the weak and strong nuclear forces) do have many things in common, especially at high energies prevalent near the big bang. However, a true unification of all forces remains elusive. Hawking was hopeful: "I think there is a good chance that the study of the early universe and the requirements of mathematical consistency will lead us to a complete unified theory by the end of the century—always presuming we don't blow ourselves up first." I imagine he still is, even though the end of the twentieth century should be shifted to the end of the twenty-first. In the meantime, we have to keep asking questions.

After all, who wouldn't want to know the mind of God?

INTRODUCTION

In this series of lectures I shall try to give an outline of what we think is the history of the universe from the big bang to black holes. In the first lecture I shall briefly review past ideas about the universe and how we got to our present picture. One might call this the history of the history of the universe.

In the second lecture I shall describe how both Newton's and Einstein's theories of gravity led to the conclusion that the universe could not be static; it had to be either expanding or contracting. This, in turn, implied that there must have been a time between ten and twenty billion years ago when the density of the universe was infinite. This is called the big bang. It would have been the beginning of the universe.

In the third lecture I shall talk about black holes. These are formed when a massive star or an even larger body collapses in on itself under its own gravitational pull. According to Einstein's general theory of relativity, anyone foolish enough to fall into a black hole will be lost forever. They will not be able to come out of the black hole again. Instead, history, as far as they are concerned, will come to a sticky end at a singularity. However, general relativity is a classical theory—that is, it doesn't take into account the uncertainty principle of quantum mechanics.

In the fourth lecture I shall describe how quantum mechanics allows energy to leak out of black holes. Black holes aren't as black as they are painted.

In the fifth lecture I shall apply quantum mechanical ideas to the big bang and the origin of the universe. This leads to the idea that space–time may be finite in extent but without boundary or edge. It would be like the surface of the Earth but with two more dimensions.

In the sixth lecture I shall show how this new boundary proposal could explain why the past is so different from the future, even though the laws of physics are time symmetric.

Finally, in the seventh lecture I shall describe how we are trying to find a unified theory that will include quantum mechanics, gravity, and all the other interactions of physics. If we achieve this, we shall really understand the universe and our position in it.

IDEAS ABOUT the UNIVERSE

As long ago as 340 B.C. Aristotle, in his book *On the Heavens*, was able to put forward two good arguments for believing that the Earth was a round ball rather than a flat plate. First, he realized that eclipses of the moon were caused by the Earth coming between the sun and the moon. The Earth's shadow on the moon was always round, which would be true only if the Earth was spherical. If the Earth had been a flat disk, the shadow would have been elongated and elliptical, unless the eclipse always occurred at a time when the sun was directly above the center of the disk.

Second, the Greeks knew from their travels that the Pole Star appeared lower in the sky when viewed in the south than it did in more northerly regions. From the difference in the apparent position of the Pole Star in Egypt and Greece, Aristotle even quoted an estimate that the distance around the Earth was four hundred thousand stadia. It is not known exactly what length a stadium was, but it may have been about two hundred yards. This would make Aristotle's estimate about twice the currently accepted figure.

The Greeks even had a third argument that the Earth must be round, for why else does one first see the sails of a ship coming over the horizon and only later see the hull? Aristotle thought that the Earth was stationary and that the sun, the moon, the planets, and the stars moved in circular orbits about the Earth. He believed this because he felt, for mystical reasons, that the Earth was the center of the universe and that circular motion was the most perfect.

Aristotle thought that the Earth was stationary and that the sun, the moon, the planets, and the stars moved in circular orbits about the Earth.

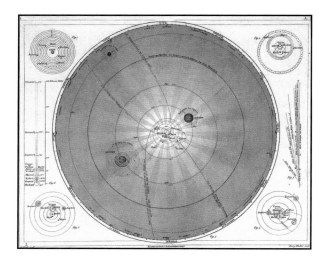

Historical artwork of the various historical cosmological models proposed to explain planetary movement. The main diagram is a heliocentric (sun-centered) model of the then six known planets, their satellites and other heavenly bodies orbiting the sun. From the Second century the dominant model had been the geocentric (Earth-centered) Ptolemaic system (upper left). This was overthrown by the heliocentric Copernican model of 1543 (lower right). An Egyptian model (lower left) and Tychonic model (upper right) attempted to keep a stationary Earth at the center of the universe. Details of planetary orbits are at left and right. From Bilder Atlas by Johann Georg Heck, 1860.

This idea was elaborated by Ptolemy in the first century A.D. into a complete cosmological model. The Earth stood at the center, surrounded by eight spheres, which carried the moon, the sun, the stars, and the five planets known at the time: Mercury, Venus, Mars, Jupiter, and Saturn. The planets themselves moved on smaller circles attached to their respective spheres in order to account for their rather complicated observed paths in the sky. The outermost sphere carried the so–called fixed stars, which always stay in the same positions relative to each other but which rotate together across the sky. What lay beyond the last sphere was never made very clear, but it certainly was not part of mankind's observable universe.

Ptolemy's model provided a reasonably accurate system for predicting the positions of heavenly bodies in the sky. But in order to predict these positions correctly, Ptolemy had to make an assumption that the moon followed a path that sometimes brought it twice as close to the Earth as at other times. And that meant that the moon had sometimes to appear twice as big as it usually does. Ptolemy was aware of this flaw but nevertheless his model was generally, although not universally, accepted. It was adopted by the Christian church as the picture of the universe that was in accordance with Scripture. It had the great advantage that it left lots of room outside the sphere of fixed stars for heaven and hell.

A much simpler model, however, was proposed in 1514 by a Polish priest, Nicholas Copernicus. At first, for fear of being accused of heresy, Copernicus published his model anonymously. His idea was that the sun was stationary at the center and that the Earth and the planets moved in circular orbits around the sun. Sadly for Copernicus, nearly a century passed before this idea was to be taken seriously. Then two astronomers—the German, Johannes Kepler, and the Italian, Galileo Galilei—started publicly to support the Copernican theory, despite the fact that the orbits it predicted did not quite match the ones observed. The death of the Aristotelian–Ptolemaic theory came in 1609. In that year Galileo started observing the night sky with a telescope, which had just been invented.

In 1609, Galileo started observing the night sky with a telescope, which had just been invented.

When he looked at the planet Jupiter, Galileo found that it was accompanied by several small satellites, or moons, which orbited around it. This implied that everything did not have to orbit directly around the Earth as Aristotle and Ptolemy had thought. It was, of course, still possible to believe that the Earth was stationary at the center of the universe, but that the moons of Jupiter moved on extremely complicated paths around the Earth, giving the appearance that they orbited Jupiter. However, Copernicus's theory was much simpler.

At the same time, Kepler had modified Copernicus's theory, suggesting that the planets moved not in circles, but in ellipses. The predictions now finally matched the observations. As far as Kepler was concerned, elliptical orbits were merely an ad hoc hypothesis—and a rather repugnant one at that because ellipses were clearly less perfect than circles. Having discovered, almost by accident, that elliptical orbits fitted the observations well, he could not reconcile with his idea that the planets were made to orbit the sun by magnetic forces.

An explanation was provided only much later, in 1687, when Newton published his *Principia Mathematica Naturalis Causae*. This was probably the most important single work ever published in the physical sciences. In it, Newton not only put forward a theory of how bodies moved in space and time, but he also developed the mathematics needed to analyze those motions. In addition, Newton postulated a law of universal gravitation. This said that each body in the universe was attracted toward every other body by a force which was stronger the more massive the bodies and the closer they were to each other. It was the same force which caused objects to fall to the ground. The story that Newton was hit on the head by an apple is almost certainly apocryphal. All Newton himself ever said was that the idea of gravity came to him as he sat in a contemplative mood, and was occasioned by the fall of an apple.

Newton went on to show that, according to his law, gravity causes the moon to move in an elliptical orbit around the Earth and causes the Earth and the planets to follow elliptical paths around the sun. The Copernican model got rid of Ptolemy's celestial spheres, and with them the idea that the universe had a natural boundary. The fixed stars did not appear to change their relative positions as the Earth went around the sun. It therefore became natural to suppose that the fixed stars were objects like our sun but much farther away. This raised a problem. Newton realized that, according to his theory of gravity, the stars should attract each other; so, it seemed they could not remain essentially motionless. Would they not all fall together at some point?

In a letter in 1691 to Richard Bentley, another leading thinker of his day, Newton argued that this would indeed happen if there were only a finite number of stars. But he reasoned that if, on the

Newton not only put forward a theory of how bodies moved in space and time, but he also developed the mathematics needed to analyze those motions.

other hand, there were an infinite number of stars distributed more or less uniformly over infinite space, this would not happen because there would not be any central point for them to fall to. This argument is an instance of the pitfalls that one can encounter when one talks about infinity.

In an infinite universe, every point can be regarded as the center because every point has an infinite number of stars on each side of it. The correct approach, it was realized only much later, is to consider the finite situation in which the stars all fall in on each other. One then asks how things change if one adds more stars roughly uniformly distributed outside this region. According to Newton's law, the extra stars would make no difference at all to the original ones, and so the stars would fall in just as fast. We can add as many stars as we like, but they will still always collapse in on themselves. We now know it is impossible to have an infinite static model of the universe in which gravity is always attractive.

It is an interesting reflection on the general climate of thought before the twentieth century that no one had suggested that the universe was expanding or contracting. It was generally accepted that either the universe had existed forever in an unchanging state or that it had been created at a finite time in the past, more or less as we observe it today. In part, this may have been due to people's tendency to believe in eternal truths as well as the comfort they found in the thought that even though they may grow old and die, the universe is unchanging.

Even those who realized that Newton's theory of gravity showed that the universe could not be static did not think to suggest that it might be expanding. Instead, they attempted to modify the theory by making the gravitational force repulsive at very large distances. This did not significantly affect their predictions of the motions of the planets. But it would allow an infinite distribution of stars to remain in equilibrium, with the attractive forces between nearby stars being balanced by the repulsive forces from those that were farther away.

Before the twentieth century no one had suggested that the universe was expanding or contracting.

While it was once thought that an infinite number of stars could remain in equilibrium, with the attractive forces between nearby stars being balanced by the repulsive forces from those that were farther away, it is now believed that such an equilibrium would be unstable. Quintuplet Cluster, one of the largest young clusters of stars inside our Milky Way galaxy, is destined to be ripped apart in just a few million years by gravitational tidal forces in the Galaxy's core. But in its brief lifetime it shines more brightly than any other star cluster in the galaxy.

However, we now believe such an equilibrium would be unstable. If the stars in some region got only slightly near each other, the attractive forces between them would become stronger and would dominate over the repulsive forces. This would mean that the stars would continue to fall toward each other. On the other hand, if the stars got a bit farther away from each other, the repulsive forces would dominate and drive them farther apart.

Another objection to an infinite static universe is normally ascribed to the German philosopher Heinrich Olbers. In fact,

View of stars in an infinite static universe.

various contemporaries of Newton had raised the problem, and the Olbers article of 1823 was not even the first to contain plausible arguments on this subject. It was, however, the first to be widely noted. The difficulty is that in an infinite static universe nearly every line of sight would end on the surface of a star. Thus one would expect that the whole sky would be as bright as the sun, even at night. Olbers's counterargument was that the light from distant stars would be dimmed by absorption by intervening matter. However, if that happened, the intervening matter would eventually heat up until it glowed as brightly as the stars.

In an infinite static universe, nearly every line of sight would end on the surface of a star.

7

The only way of avoiding the conclusion that the whole of the night sky should be as bright as the surface of the sun would be if the stars had not been shining forever, but had turned on at some finite time in the past. In that case, the absorbing matter might not have heated up yet, or the light from distant stars might not yet have reached us. And that brings us to the question of what could have caused the stars to have turned on in the first place.

THE BEGINNING OF THE UNIVERSE

The beginning of the universe had, of course, been discussed for a long time. According to a number of early cosmologies in the Jewish/Christian/Muslim tradition, the universe started at a finite and not very distant time in the past. One argument for such a beginning was the feeling that it was necessary to have a first cause to explain the existence of the universe.

Another argument was put forward by St. Augustine in his book, *The City of God*. He pointed out that civilization is progressing, and we remember who performed this deed or developed that technique. Thus man, and so also perhaps the universe, could not have been around all that long. For otherwise we would have already progressed more than we have.

St. Augustine accepted a date of about 5000 B.C. for the creation of the universe according to the book of Genesis. It is interesting that this is not so far from the end of the last Ice Age, about 10,000 B.C., which is when civilization really began. Aristotle and most of the other Greek philosophers, on the other hand, did not like the idea of a creation because it made too much of divine intervention. They believed, therefore, that the human race and the world around it had existed, and would exist, forever. They had already considered the argument about progress, described earlier, and answered it by saying that there had been periodic floods or other disasters that repeatedly set the human race right back to the beginning of civilization.

When most people believed in an essentially static and unchanging universe, the question of whether or not it had a beginning was really one of metaphysics or theology. One could account for what was observed either way. Either the universe had existed forever, or it was set in motion at some finite time in such a manner as to look as though it had existed forever. But in 1929, Edwin Hubble made the landmark observation that wherever you look, distant stars are moving rapidly away from us. In other words, the universe is expanding. This means that at earlier times objects would have been closer together.

In fact, it seemed that there was a time about ten or twenty thousand million years ago when they were all at exactly the same place. This discovery finally brought the question of the beginning of the universe into the realm of science. Hubble's observations suggested that there was a time called the big bang when the universe was infinitesimally small and, therefore, infinitely dense. If there were events earlier than this time, then they could not affect what happens at the present time. Their existence can be ignored because it would have no observational consequences.

One may say that time had a beginning at the big bang, in the sense that earlier times simply could not be defined. It should be emphasized that this beginning in time is very different from those that had been considered previously. In an unchanging universe, a beginning in time is something that has to be imposed by some being outside the universe. There is no physical necessity for a beginning. One can imagine that God created the universe at literally any time in the past. On the other hand, if the universe is expanding, there may be physical reasons why there had to be a beginning. One could still believe that God created the universe at the instant of the big bang. He could even have created it at a later time in just such a way as to make it look as though there had been a big bang. But it would be meaningless to suppose that it was created before the big bang. An expanding universe does not preclude a creator, but it does place limits on when He might have carried out his job.

THE EXPANDING UNIVERSE

O ur sun and the nearby stars are all part of a vast collection of stars called the Milky Way galaxy. For a long time it was thought that this was the whole universe. It was only in 1924 that the American astronomer Edwin Hubble demonstrated that ours was not the only galaxy. There were, in fact, many others, with vast tracks of empty space between them. In order to prove this he needed to determine the distances to these other galaxies. We can determine the distance of nearby stars by observing how they change position as the Earth goes around the sun. But other galaxies are so far away that, unlike nearby stars, they really do appear fixed. Hubble was forced, therefore, to use indirect methods to measure the distances.

Now the apparent brightness of a star depends on two factors—luminosity and how far it is from us. For nearby stars we can measure both their apparent brightness and their distance, so we can work out their luminosity. Conversely, if we knew the luminosity of stars in other galaxies, we could work out their distance by measuring their apparent brightness. Hubble argued that there were certain types of stars that always had the same luminosity when they were near enough for us to measure. If, therefore, we found such stars in another galaxy, we could assume that they had the same luminosity. Thus, we could calculate the distance to that galaxy. If we could do this for a number of stars in the same galaxy, and our calculations always gave the same distance, we could be fairly confident of our estimate. In this way, Edwin Hubble worked out the distances to nine different galaxies.

We can determine the distance of nearby stars by observing how they change position as the Earth goes around the sun.

11

Located some 13 million light-years from Earth, NGC 4214 is currently forming clusters of new stars from its interstellar gas and dust. In this Hubble image, we can see a sequence of steps in the formation and evolution of stars and star clusters. The youngest of these star clusters are located at the lower right of the picture, where they appear as about half a dozen bright clumps of glowing gas. Here the young, hot stars have a whitish to bluish color in the Hubble image, because of their high surface temperatures, ranging from 10,000 up to about 50,000 degrees Celsius. Moving to the lower left from the youngest clusters, we find an older star cluster. The most spectacular feature in the Hubble picture, lying near the center of NGC 4214, is a cluster of hundreds of massive blue stars, each of them more than ten thousand times brighter than our own sun.

We now know that our galaxy is only one of some hundred thousand million that can be seen using modern telescopes, each galaxy itself containing some hundred thousand million stars. We live in a galaxy that is about one hundred thousand light-years across and is slowly rotating; the stars in its spiral arms orbit around its center about once every hundred million years. Our sun is just an ordinary, average-sized, yellow star, near the outer edge of one of the spiral arms. We have certainly come a long way since Aristotle and Ptolemy, when we thought that the Earth was the center of the universe.

Stars are so far away that they appear to us to be just pinpoints of light. We cannot determine their size or shape. So how can we tell different types of stars apart? For the vast majority of stars, there is only one correct characteristic feature that we can observe—the color of their light. Newton discovered that if light from the sun passes through a prism, it breaks up into its component colors—its spectrum—like in a rainbow. By focusing a telescope on an individual star or galaxy, one can similarly observe the spectrum of the light from that star or galaxy.

Different stars have different spectra, but the relative brightness of the different colors is always exactly what one would expect to find in the light emitted by an object that is glowing red hot. This means that we can tell a star's temperature from the spectrum of its light. Moreover, we find that certain very specific colors are missing from stars' spectra, and these missing colors may vary from star to star. We know that each chemical element absorbs the characteristic set of very specific colors. Thus, by matching each of those which are missing from a star's spectrum,

One of the deepest images of the sky taken to date with NASA's Hubble Space Telescope reveals a population of faint blue galaxies which turn out to be the most common class of objects in the universe. Their distances are estimated to range from three to eight billion light-years, meaning than they were abundant when the universe was a fraction of its present age, but are rare or harder to find today because they have faded or self-destructed. Deciphering the formation and evolution of these blue dwarf galaxies may provide new clues to understanding the process of galaxy evolution, including the formation of our Milky Way galaxy. The galaxies are blue because they are undergoing episodes of intense star-formation which produce a lot of young, hot, and blue stars.

we can determine exactly which elements are present in the star's atmosphere.

In the 1920s, when astronomers began to look at the spectra of stars in other galaxies, they found something most peculiar: There were the same characteristic sets of missing colors as for stars in our own galaxy, but they were all shifted by the same relative amount toward the red end of the spectrum. The only reasonable explanation of this was that the galaxies were moving away from us, and the frequency of the light waves from them was being reduced, or red-shifted, by the Doppler effect. Listen to a car passing on the road. As the car is approaching, its engine sounds at a higher pitch, corresponding to a higher frequency of sound waves; and when it passes and goes away, it sounds at a lower pitch. The behavior of light or radial waves is similar. Indeed, the police made use of the Doppler effect to measure the speed of cars by measuring the frequency of pulses of radio waves reflected off them.

In the years following his proof of the existence of other galaxies, Hubble spent his time cataloging their distances and observing their spectra. At that time most people expected the galaxies to be moving around quite randomly, and so expected to find as many spectra which were blue-shifted as ones which were red–shifted. It was quite a surprise, therefore, to find that the galaxies all appeared red-shifted. Every single one was moving away from us. More surprising still was the result which Hubble published in 1929: Even the size of the galaxy's red shift was not random, but was directly proportional to the galaxy's distance from us. Or, in other words, the farther a galaxy was, the faster it was moving away. And that meant that the universe could not be static, as everyone previously thought, but was in fact expanding. The distance between the different galaxies was growing all the time.

The discovery that the universe was expanding was one of the great intellectual revolutions of the twentieth century. With

Even the size of the galaxy's red shift was not random, but was directly proportional to the galaxy's distance from us. Or, in other words, the farther a galaxy was, the faster it was moving away.

hindsight, it is easy to wonder why no one had thought of it before. Newton and others should have realized that a static universe would soon start to contract under the influence of gravity. But suppose that, instead of being static, the universe was expanding. If it was expanding fairly slowly, the force of gravity would cause it eventually to stop expanding and then to start contracting. However, if it was expanding at more than a certain critical rate, gravity would never be strong enough to stop it, and the universe would continue to expand forever. This is a bit like what happens when one fires a rocket upward from the surface of the Earth. If it has a fairly low speed, gravity will eventually stop the rocket, and it will start falling back. On the other hand, if the rocket has more than a certain critical speed—about seven miles a second—gravity will not be strong enough to pull it back, so it will keep going away from the Earth forever.

This behavior of the universe could have been predicted from Newton's theory of gravity at any time in the nineteenth, the eighteenth, or even the late seventeenth centuries. Yet so strong was the belief in a static universe that it persisted into the early twentieth century. Even when Einstein formulated the general theory of relativity in 1915, he was sure that the universe had to be static. He therefore modified his theory to make this possible, introducing a so-called cosmological constant into his equations. This was a new "antigravity" force, which, unlike other forces, did not come from any particular source, but was built into the very fabric of space-time. His cosmological constant gave space-time an inbuilt tendency to expand, and this could be made to exactly balance the attraction of all the matter in the universe so that a static universe would result.

Only one man, it seems, was willing to take general relativity at face value. While Einstein and other physicists were looking for ways of avoiding general relativity's prediction of a nonstatic universe, the Russian physicist Alexander Friedmann instead set about explaining it.

Even when Einstein formulated the general theory of relativity in 1915, he was sure that the universe had to be static.

To determine whether or not the universe will stop expanding and eventually begin to contract or whether it will continue to expand forever, one could compare it to a rocket leaving Earth. If the rocket has a fairly low speed, gravity will eventually stop the rocket, and it will start falling back to Earth. If the rocket has more than the critical speed—about seven miles a second—gravity will not be strong enough to pull it back, so it will keep going away from the Earth forever. NASA successfully launched more than two hundred Earth-orbiting satellites, including Goddard's eighth Orbiting Solar Observatory aboard this Delta rocket on June 21, 1975, at Cape Canaveral, Florida.

THE FRIEDMANN MODELS

The equations of general relativity, which determined how the universe evolves in time, are too complicated to solve in detail. So what Friedmann did, instead, was to make two very simple assumptions about the universe: that the universe looks identical in whichever direction we look, and that this would also be true if we were observing the universe from anywhere else. On the basis of general relativity and these two assumptions, Friedmann showed that we should not expect the universe to be static. In fact, in 1922, several years before Edwin Hubble's discovery, Friedmann predicted exactly what Hubble found.

The assumption that the universe looks the same in every direction is clearly not true in reality. For example, the other stars in our galaxy form a distinct band of light across the night sky called the Milky Way. But if we look at distant galaxies, there seems to be more or less the same number of them in each direction. So the universe does seem to be roughly the same in every direction, provided one views it on a large scale compared to the distance between galaxies.

For a long time this was sufficient justification for Friedmann's assumption—as a rough approximation to the real universe. But more recently a lucky accident uncovered the fact that Friedmann's assumption is in fact a remarkably accurate description of our universe. In 1965, two American physicists, Arno Penzias and Robert Wilson, were working at the Bell Labs in New Jersey on the design of a very sensitive microwave detector for communicating with orbiting satellites. They were worried when they found that their detector was picking up more noise than it ought to, and that the noise did not appear to be coming from any particular direction. First they looked for bird droppings on their detector and checked for other possible malfunctions, but soon ruled these out. They knew that any noise from within the atmosphere would be stronger when the detector is not pointing straight up than when it is, because the atmosphere appears thicker when looking at an angle to the vertical.

Friedmann did make two very simple assumptions about the universe: that the universe looks identical in whichever direction we look, and that this would also be true if we were observing the universe from anywhere else.

The Chandra X-Ray observatory (CXO) has made a stunning, high-energy panorama of the central regions of our Milky Way Galaxy. This 400 by 900-light-year mosaic of several CXO images reveals hundreds of white dwarf stars, neutron stars, and black holes bathed in an incandescent fog of multimillion-degree gas.

The extra noise was the same whichever direction the detector pointed, so it must have come from outside the atmosphere.

The extra noise was the same whichever direction the detector pointed, so it must have come from outside the atmosphere. It was also the same day and night throughout the year, even though the Earth was rotating on its axis and orbiting around the sun. This showed that the radiation must come from beyond the solar system, and even from beyond the galaxy, as otherwise it would vary as the Earth pointed the detector in different directions.

In fact, we know that the radiation must have traveled to us across most of the observable universe. Since it appears to be the same in different directions, the universe must also be the same in every direction, at least on a large scale. We now know that whichever direction we look in, this noise never varies by more than one part in ten thousand. So Penzias and Wilson had unwittingly stumbled across a remarkably accurate confirmation of Friedmann's first assumption.

At roughly the same time, two American physicists at nearby Princeton University, Bob Dicke and Jim Peebles, were also taking an interest in microwaves. They were working on a suggestion made by George Gamow, once a student of Alexander Friedmann, that the early universe should have been very hot and dense, glowing white hot. Dicke and Peebles argued that we should still be able to see this glowing, because light from very distant parts of the early universe would only just be reaching us now. However, the expansion of the universe meant that this light should be so greatly red-shifted that it would appear to us now as microwave radiation. Dicke and Peebles were looking for this radiation when Penzias and Wilson heard about their work and realized that they had already found it. For this, Penzias and Wilson were awarded the Nobel Prize in 1978, which seems a bit hard on Dicke and Peebles.

Now at first sight, all this evidence that the universe looks the same whichever direction we look in might seem to suggest there is something special about our place in the universe. In particular, it might seem that if we observe all other galaxies to be moving away from us, then we must be at the center of the universe. There is, however, an alternative explanation: The universe might also look the same in every direction as seen from any other galaxy. This, as we have seen, was Friedmann's second assumption.

We have no scientific evidence for or against this assumption. We believe it only on grounds of modesty. It would be most remarkable if the universe looked the same in every direction around us, but not around other points in the universe. In Friedmann's model, all the galaxies are moving directly away from each other. The situation is rather like steadily blowing up a balloon which has a number of spots painted on it. As the balloon expands, the distance between any two spots increases, but there is no spot that can be said to be the center of the expansion. Moreover, the farther apart the spots are, the faster they will be moving apart. Similarly, in Friedmann's model the speed at which any two galaxies are moving apart is proportional to the distance between them. So it predicted that the red shift of a galaxy should be directly proportional to its distance from us, exactly as Hubble found.

Despite the success of his model and his prediction of Hubble's observations, Friedmann's work remained largely unknown in the West. It became known only after similar models

It might seem that if we observe all other galaxies to be moving away from us, then we must be at the center of the universe.

were discovered in 1935 by the American physicist Howard Robertson and the British mathematician Arthur Walker, in response to Hubble's discovery of the uniform expansion of the universe.

Although Friedmann found only one, there are in fact three different kinds of models that obey Friedmann's two fundamental assumptions. In the first kind—which Friedmann found—the universe is expanding so sufficiently slowly that the gravitational attraction between the different galaxies causes the expansion to slow down and eventually to stop. The galaxies then start to move toward each other and the universe contracts. The distance between two neighboring galaxies starts at zero, increases to a maximum, and then decreases back down to zero again.

In the second kind of solution, the universe is expanding so rapidly that the gravitational attraction can never stop it, though it does slow it down a bit. The separation between neighboring galaxies in this model starts at zero, and eventually the galaxies are moving apart at a steady speed.

Finally, there is a third kind of solution, in which the universe is expanding only just fast enough to avoid recollapse. In this case the separation also starts at zero, and increases forever. However, the speed at which the galaxies are moving apart gets smaller and smaller, although it never quite reaches zero.

A remarkable feature of the first kind of Friedmann model is that the universe is not infinite in space, but neither does space have any boundary. Gravity is so strong that space is bent round onto itself, making it rather like the surface of the Earth. If one keeps traveling in a certain direction on the surface of the Earth, one never comes up against an impassable barrier or falls over the edge, but eventually comes back to where one started. Space, in the first Friedmann model, is just like this, but with three dimensions instead of two for the Earth's surface. The fourth dimension—time—is also finite in extent, but it is like a line with two ends or boundaries, a beginning and an end. We shall see later that when one combines general relativity with the

A remarkable feature of the first kind of Friedmann model is that the universe is not infinite in space, but neither does space have any boundary.

uncertainty principle of quantum mechanics, it is possible for both space and time to be finite without any edges or boundaries. The idea that one could go right around the universe and end up where one started makes good science fiction, but it doesn't have much practical significance because it can be shown that the universe would recollapse to zero size before one could get round. You would need to travel faster than light in order to end up where you started before the universe came to an end—and that is not allowed.

But which Friedmann model describes our universe? Will the universe eventually stop expanding and start contracting, or will it expand forever? To answer this question we need to know the present rate of expansion of the universe and its present average density. If the density is less than a certain critical value, determined by the rate of expansion, the gravitational attraction will be too weak to halt the expansion. If the density is greater than the critical value, gravity will stop the expansion at some time in the future and cause the universe to recollapse.

We can determine the present rate of expansion by measuring the velocities at which other galaxies are moving away from us, using the Doppler effect. This can be done very accurately. However, the distances to the galaxies are not very well known because we can only measure them indirectly. So all we know is that the universe is expanding by between 5 percent and 10 percent every thousand million years. However, our uncertainty about the present average density of the universe is even greater.

If we add up the masses of all the stars that we can see in our galaxy and other galaxies, the total is less than one-hundredth of the amount required to halt the expansion of the universe, even in the lowest estimate of the rate of expansion. But we know that our galaxy and other galaxies must contain a large amount of dark matter which we cannot see directly, but which we know must be there because of the influence of its gravitational attraction on the orbits of stars and gas in the galaxies. Moreover,

We can determine the present rate of expansion by measuring the velocities at which other galaxies are moving away from us, using the Doppler effect.

If the universe is

going to recollapse,

it won't do so for

at least another ten

thousand million

years.

most galaxies are found in clusters, and we can similarly infer the presence of yet more dark matter in between the galaxies in these clusters by its effect on the motion of the galaxies. When we add up all this dark matter, we still get only about one-tenth of the amount required to halt the expansion. However, there might be some other form of matter which we have not yet detected and which might still raise the average density of the universe up to the critical value needed to halt the expansion.

The present evidence, therefore, suggests that the universe will probably expand forever. But don't bank on it. All we can really be sure of is that even if the universe is going to recollapse, it won't do so for at least another ten thousand million years, since it has already been expanding for at least that long. This should not unduly worry us since by that time, unless we have colonies beyond the solar system, mankind will long since have died out, extinguished along with the death of our sun.

THE BIG BANG

All of the Friedmann solutions have the feature that at some time in the past, between ten and twenty thousand million years ago, the distance between neighboring galaxies must have been zero. At that time, which we call the big bang, the density of the universe and the curvature of space-time would have been infinite. This means that the general theory of relativity—on which Friedmann's solutions are based—predicts that there is a singular point in the universe.

All our theories of science are formulated on the assumption that space–time is smooth and nearly flat, so they would all break down at the big bang singularity, where the curvature of space–time is infinite. This means that even if there were events before the big bang, one could not use them to determine what would happen afterward, because predictability would break down at the big bang. Correspondingly, if we know only what has happened since the big bang, we could not determine what happened beforehand. As far as we are concerned, events before the big bang can have no consequences, so they should not form part of a scientific model of the universe. We should therefore cut them out of the model and say that time had a beginning at the big bang.

Many people do not like the idea that time has a beginning, probably because it smacks of divine intervention. (The Catholic church, on the other hand, had seized on the big bang model and in 1951 officially pronounced it to be in accordance with the Bible.) There were a number of attempts to avoid the conclusion that there had been a big bang. The proposal that gained widest support was called the steady state theory. It was suggested in 1948 by two refugees from Nazi–occupied Austria, Hermann Bondi and Thomas Gold, together with the Briton Fred Hoyle, who had worked with them on the development of radar during the war. The idea was that as the galaxies moved away from each other, new galaxies were continually forming in the gaps in between, from new matter that was being continually created. The universe would therefore look roughly the same at all times as well as at all points of space.

The steady state theory required a modification of general relativity to allow for the continual creation of matter, but the rate that was involved was so low—about one particle per cubic kilometer per year—that it was not in conflict with experiment. The theory was a good scientific theory, in the sense that it was simple and it made definite predictions that could be tested by observation. One of these predictions was that the number of galaxies or similar objects in any given volume of space should be the same wherever and whenever we look in the universe.

In the late 1950s and early 1960s, a survey of sources of radio waves from outer space was carried out at Cambridge by a group of astronomers led by Martin Ryle. The Cambridge group showed that most of these radio sources must lie outside our

There were a number of attempts to avoid the conclusion that there had been a big bang.

galaxy, and also that there were many more weak sources than strong ones. They interpreted the weak sources as being the more distant ones, and the stronger ones as being near. Then there appeared to be fewer sources per unit volume of space for the nearby sources than for the distant ones.

This could have meant that we were at the center of a great region in the universe in which the sources were fewer than elsewhere. Alternatively, it could have meant that the sources were more numerous in the past, at the time that the radio waves left on their journey to us, than they are now. Either explanation contradicted the predictions of the steady state theory. Moreover, the discovery of the microwave radiation by Penzias and Wilson in 1965 also indicated that the universe must have been much denser in the past. The steady state theory therefore had regretfully to be abandoned.

Another attempt to avoid the conclusion that there must have been a big bang and, therefore, a beginning of time, was made by two Russian scientists, Evgenii Lifshitz and Isaac Khalatnikov, in 1963. They suggested that the big bang might be a peculiarity of Friedmann's models alone, which after all were only approximations to the real universe. Perhaps, of all the models that were roughly like the real universe, only Friedmann's would contain a big bang singularity. In Friedmann's models, the galaxies are all moving directly away from each other. So it is not surprising that at some time in the past they were all at the same place. In the real universe, however, the galaxies are not just moving directly away from each other—they also have small sideways velocities. So in reality they need never have been all at exactly the same place, only very close together. Perhaps, then, the current expanding universe resulted not from a big bang singularity, but from an earlier contracting phase; as the universe had collapsed, the particles in it might not have all collided, but they might have flown past and then away from each other, producing the present expansion of the universe. How then

Galaxies are not just moving directly away from each other—they also have small sideways velocities.

could we tell whether the real universe should have started out with a big bang?

What Lifshitz and Khalatnikov did was to study models of the universe which were roughly like Friedmann's models but which took account of the irregularities and random velocities of galaxies in the real universe. They showed that such models could start with a big bang, even though the galaxies were no longer always moving directly away from each other. But they claimed that this was still only possible in certain exceptional models in which the galaxies were all moving in just the right way. They argued that since there seemed to be infinitely more Friedmann-like models without a big bang singularity than there were with one, we should conclude that it was very unlikely that there had been a big bang. They later realized, however, that there was a much more general class of Friedmann-like models which did have singularities, and in which the galaxies did not have to be moving in any special way. They therefore withdrew their claim in 1970.

The work of Lifshitz and Khalatnikov was valuable because it showed that the universe could have had a singularity—a big bang—if the general theory of relativity was correct. However, it did not resolve the crucial question: Does general relativity predict that our universe should have the big bang, a beginning of time? The answer to this came out of a completely different approach started by a British physicist, Roger Penrose, in 1965. He used the way light cones behave in general relativity, and the fact that gravity is always attractive, to show that a star that collapses under its own gravity is trapped in a region whose boundary eventually shrinks to zero size. This means that all the matter in the star will be compressed into a region of zero volume, so the density of matter and the curvature of space-time become infinite. In other words, one has a singularity contained within a region of space-time known as a black hole.

Roger Penrose used the way light cones behave in general relativity, and the fact that gravity is always attractive, to show that a star that collapses under its own gravity is trapped in a region whose boundary eventually shrinks to zero size.

There must

have been a big

bang singularity

provided only that

general relativity

is correct and

that the universe

contains as much

matter as we

observe.

At first sight, Penrose's result didn't have anything to say about the question of whether there was a big bang singularity in the past. However, at the time that Penrose produced his theorem, I was a research student desperately looking for a problem with which to complete my Ph.D. thesis. I realized that if one reversed the direction of time in Penrose's theorem so that the collapse became an expansion, the conditions of his theorem would still hold, provided the universe were roughly like a Friedmann model on large scales at the present time. Penrose's theorem had shown that any collapsing star must end in a singularity; the time-reversed argument showed that any Friedmann-like expanding universe must have begun with a singularity. For technical reasons, Penrose's theorem required that the universe be infinite in space. So I could use it to prove that there should be a singularity only if the universe was expanding fast enough to avoid collapsing again, because only that Friedmann model was infinite in space.

During the next few years I developed new mathematical techniques to remove this and other technical conditions from the theorems that proved that singularities must occur. The final result was a joint paper by Penrose and myself in 1970, which proved that there must have been a big bang singularity provided only that general relativity is correct and that the universe contains as much matter as we observe.

There was a lot of opposition to our work, partly from the Russians, who followed the party line laid down by Lifshitz and Khalatnikov, and partly from people who felt that the whole idea of singularities was repugnant and spoiled the beauty of Einstein's theory. However, one cannot really argue with the mathematical theorem. So it is now generally accepted that the universe must have a beginning.

BLACK HOLES

The term *black hole* is of very recent origin. It was coined in 1969 by the American scientist John Wheeler as a graphic description of an idea that goes back at least two hundred years. At that time there were two theories about light. One was that it was composed of particles; the other was that it was made of waves. We now know that really both theories are correct. By the wave/particle duality of quantum mechanics, light can be regarded as both a wave and a particle. Under the theory that light was made up of waves, it was not clear how it would respond to gravity. But if light were composed of particles, one might expect them to be affected by gravity in the same way that cannonballs, rockets, and planets are.

On this assumption, a Cambridge don, John Michell, wrote a paper in 1783 in the *Philosophical Transactions of the Royal Society of London.* In it, he pointed out that a star that was sufficiently massive and compact would have such a strong gravitational field that light could not escape. Any light emitted from the surface of the star would be dragged back by the star's gravitational attraction before it could get very far. Michell suggested that there might be a large number of stars like this. Although we would not be able to see them because the light from them would not reach us, we would still feel their gravitational attraction. Such objects are what we now call black holes, because that is what they are—black voids in space.

A similar suggestion was made a few years later by the French scientist the Marquis de Laplace, apparently independently of Michell. Interestingly enough, he included it in only the first and second editions of his book, *The System of the World,* and left it out of later editions; perhaps he decided that it was a crazy idea.

A star that was sufficiently massive and compact would have such a strong gravitational field that light could not escape.

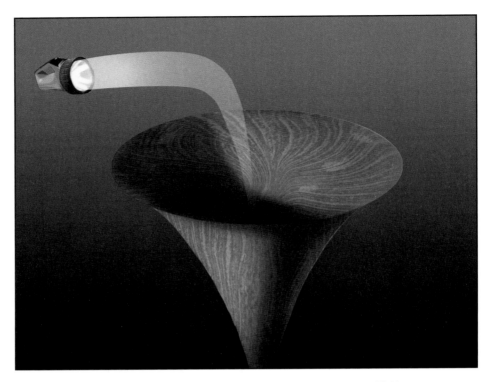

A star that is sufficiently massive and compact has such a strong gravitational field that light is dragged in and cannot escape. This kind of star is called a black hole.

A consistent theory

of how gravity

affects light did not

come until Einstein

proposed general

relativity in 1915.

In fact, it is not really consistent to treat light like cannonballs in Newton's theory of gravity because the speed of light is fixed. A cannonball fired upward from the Earth will be slowed down by gravity and will eventually stop and fall back. A photon, however, must continue upward at a constant speed. How, then, can Newtonian gravity affect light? A consistent theory of how gravity affects light did not come until Einstein proposed general relativity in 1915; and even then it was a long time before the implications of the theory for massive stars were worked out.

To understand how a black hole might be formed, we first need an understanding of the life cycle of a star. A star is formed when a large amount of gas, mostly hydrogen, starts to collapse in on itself due to its gravitational attraction. As it contracts, the atoms of the gas collide with each other more and more frequently and at greater and greater speeds—the gas heats up. Eventually the gas will be so hot that when the hydrogen atoms collide they no longer bounce off each other but instead merge with each other to form helium atoms. The heat released in this reaction, which is like a controlled hydrogen bomb, is what makes the stars shine. This additional heat also increases the pressure of the gas until it is sufficient to balance the gravitational attraction, and the gas stops contracting. It is a bit like a balloon where there is a balance between the pressure of the air inside, which is trying to make the balloon expand, and the tension in the rubber, which is trying to make the balloon smaller.

The stars will remain stable like this for a long time, with the heat from the nuclear reactions balancing the gravitational attraction. Eventually, however, the star will run out of its hydrogen and other nuclear fuels. And paradoxically, the more fuel a star starts off with, the sooner it runs out. This is because the

Artwork of an Einstein ring, formed when two massive objects are perfectly aligned with each other as seen from Earth. Here, a black hole (center) is between Earth and a galaxy. Light from the distant galaxy is bent around the black hole by the latter's immense gravitational field, forming a ring of light. The phenomenon is known as gravitational lensing. The idea that light could be bent by gravity was put forward by Albert Einstein in his general theory of relativity (1915). Several examples of gravitational lenses have been discovered in recent years.

The more massive the star is, the hotter it needs to be to balance its gravitational attraction.

more massive the star is, the hotter it needs to be to balance its gravitational attraction. And the hotter it is, the faster it will use up its fuel. Our sun has probably got enough fuel for another five thousand million years or so, but more massive stars can use up their fuel in as little as one hundred million years, much less than the age of the universe. When the star runs out of fuel, it will start to cool off and so to contract. What might happen to it then was only first understood at the end of the 1920s.

In 1928 an Indian graduate student named Subrahmanyan Chandrasekhar set sail for England to study at Cambridge with

This montage of 1999 Hubble images of a mysterious, complex structure within the Carina Nebula shows numerous small dark globules than may be in the process of collapsing to form new stars. Two striking, large, sharp-edged dust clouds are located near the bottom center and upper left edges of the image. These large dark clouds may eventually evaporate, or if there are sufficiently dense condensations within them, give birth to small star clusters. The Carina Nebula, with an overall diameter of more than two hundred light-years, is one of the outstanding features of the Southern Hemisphere portion of the Milky Way.

NASA's Hubble Space Telescope captures various stages of the life cycle of stars in one single true-color view. To the upper left of center is the evolved blue supergiant called Sher 25. Near the center of the view is a so-called starburst cluster dominated by young, hot Wolf-Rayet stars and early O-Type stars. A torrent of ionizing radiation and fast stellar winds from these massive stars has blown a large cavity around the cluster. Dark clouds at the upper right are so-called Bok Globules, which are probably in an earlier stage of star formation.

the British astronomer Sir Arthur Eddington. Eddington was an expert on general relativity. There is a story that a journalist told Eddington in the early 1920s that he had heard there were only three people in the world who understood general relativity. Eddington replied, "I am trying to think who the third person is."

During his voyage from India, Chandrasekhar worked out how big a star could be and still separate itself against its own gravity after it had used up all its fuel. The idea was this: When the star becomes small, the matter particles get very near each other. But the Pauli exclusion principle says that two matter particles cannot have both the same position and the same velocity. The matter particles must therefore have very different velocities. This makes them move away from each other, and

31

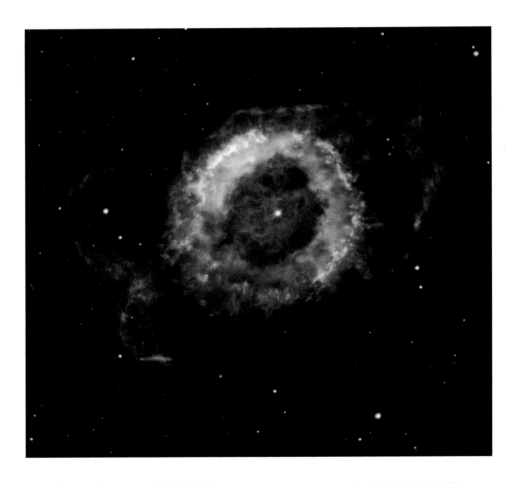

The planetary nebula NGC 6369 is known to amateur astronomers as the "Little Ghost Nebula," because it appears as a small, ghostly cloud surrounding the faint, dying central star. When a star with a mass similar to that of our own sun nears the end of its lifetime, it expands in size to become a red giant. The red-giant stage ends when the star expels its outer layers into space, producing a faintly glowing nebula. The remnant stellar core in the center is now sending out a flood of ultraviolet (UV) light into the surrounding gas. Far outside the main body of the nebula, one can see fainter wisps of gas that were lost from the star at the beginning of the ejection process. Our own sun may eject a similar nebula, but not for another 5 billion years. The gas will expand away from the star at about 15 miles per second, dissipating into interstellar space after some ten thousand years. After that, the remnant stellar ember in the center will gradually cool off for billions of years as a tiny white dwarf star, and eventually wink out.

so tends to make the star expand. A star can therefore maintain itself at a constant radius by a balance between the attraction of gravity and the repulsion that arises from the exclusion principle, just as earlier in its life the gravity was balanced by the heat.

Chandrasekhar realized, however, that there is a limit to the repulsion that the exclusion principle can provide. The theory of relativity limits the maximum difference in the velocities of the matter particles in the star to the speed of light. This meant that when the star got sufficiently dense, the repulsion caused by the exclusion principle would be less than the attraction of gravity. Chandrasekhar calculated that a cold star of more than about one and a half times the mass of the sun would not be able to support itself against its own gravity. This mass is now known as the *Chandrasekhar limit*.

This had serious implications for the ultimate fate of massive stars. If a star's mass is less than the Chandrasekhar limit, it can eventually stop contracting and settle down to a possible final state as a *white dwarf* with a radius of a few thousand miles and a density of hundreds of tons per cubic inch. A white dwarf is supported by the exclusion principle repulsion between the electrons in its matter. We observe a large number of these white dwarf stars. One of the first to be discovered is the star that is orbiting around Sirius, the brightest star in the night sky.

It was also realized that there was another possible final state for a star also with a limiting mass of about one or two times the mass of the sun, but much smaller than even the white dwarf. These stars would be supported by the exclusion principle repulsion between the neutrons and protons, rather than between the electrons. They were therefore called neutron stars. They would have had a radius of only ten miles or so and a density of hundreds of millions of tons per cubic inch. At the time they were first predicted, there was no way that neutron stars could have been observed, and they were not detected until much later.

Chandrasekhar calculated that a cold star of more than about one and a half times the mass of the sun would not be able to support itself against its own gravity.

The hostility of

other scientists,

particularly of

Eddington, his

former teacher

and the leading

authority on

the structure of

stars, persuaded

Chandrasekhar

to abandon this

line of work and

turn instead to

other problems in

astronomy.

Stars with masses above the Chandrasekhar limit, on the other hand, have a big problem when they come to the end of their fuel. In some cases they may explode or manage to throw off enough matter to reduce their mass below the limit, but it was difficult to believe that this always happened, no matter how big the star. How would it know that it had to lose weight? And even if every star managed to lose enough mass, what would happen if you added more mass to a white dwarf or neutron star to take it over the limit? Would it collapse to infinite density?

Eddington was shocked by the implications of this and refused to believe Chandrasekhar's result. He thought it was simply not possible that a star could collapse to a point. This was the view of most scientists. Einstein himself wrote a paper in which he claimed that stars would not shrink to zero size. The hostility of other scientists, particularly of Eddington, his former teacher and the leading authority on the structure of stars, persuaded Chandrasekhar to abandon this line of work and turn instead to other problems in astronomy. However, when he was awarded the Nobel Prize in 1983, it was, at least in part, for his early work on the limiting mass of cold stars.

Chandrasekhar had shown that the exclusion principle could not halt the collapse of a star more massive than the Chandrasekhar limit. But the problem of understanding what would happen to such a star, according to general relativity, was not solved until 1939 by a young American, Robert Oppenheimer. His result, however, suggested that there would be no observational consequences that could be detected by the telescopes of the day. Then the war intervened, and Oppenheimer himself became closely involved in the atom bomb project. And after the war the problem of gravitational collapse was largely forgotten as most scientists were then interested in what happens on the scale of the atom and its nucleus. In the 1960s, however, interest in the large-scale problems of astronomy and cosmology was revived by a great increase in the number and range of

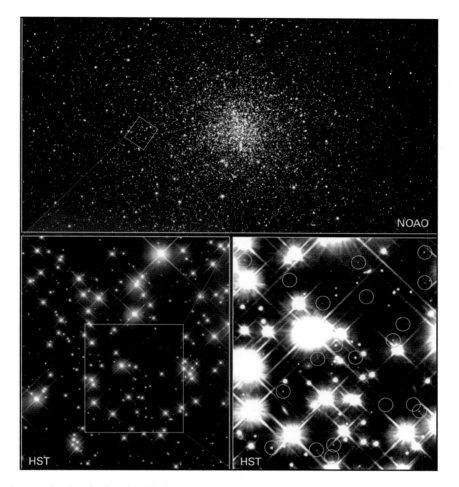

If a star's mass is less than the Chandrasekhar limit, it can eventually stop contracting and settle down to a possible final state as a white dwarf. We observe these white dwarf stars in our Milky Way galaxy: Located in the globular cluster M4, these small, burned-out stars are about 12 to 13 billion years old. By adding the one billion years it took the cluster to form after the big bang, astronomers found that the age of the white dwarfs agrees with previous estimates that the universe is 13 to 14 billion years old. In the top panel, a ground-based observatory snapped a panoramic view of the entire cluster, which contains several hundred thousand stars within a volume of 10 to 30 light-years across (1995). The box at left indicates the small region of the cluster observed by the Hubble telescope. A sampling of an even smaller region is shown at bottom right. In this smaller region, Hubble pinpointed a number of faint white dwarfs. The blue circles indicated the dwarfs. It took nearly eight days of exposure time over a 67-day period to find these extremely faint stars.

This stellar swarm is M80 (NGC 6093), one of the densest of the 147 known globular star clusters in the Milky Way galaxy. Located about 28,000 light-years from Earth, M80 contains hundreds of thousands of stars, all held together by their mutual gravitational attraction. Globular clusters are particularly useful for studying stellar evolution, since all of the stars in the clusters have the same age (about 15 billion years), but cover a range of stellar masses. Every star visible in this image is either more highly evolved than, or in a few rare cases more massive than, our own sun. Especially obvious are the bright red giants, which are stars similar to the sun in mass that are nearing the ends of their lives.

astronomical observations brought about by the application of modern technology. Oppenheimer's work was then rediscovered and extended by a number of people.

The picture that we now have from Oppenheimer's work is as follows: The gravitational field of the star changes the paths of light rays in space–time from what they would have been had the star not been present. The light cones, which indicate the paths followed in space and time by flashes of light emitted from their tips, are bent slightly inward near the surface of the star. This can be seen in the bending of light from distant stars that is observed during an eclipse of the sun. As the star contracts, the gravitational field at its surface gets stronger and the light cones get bent inward more. This makes it more difficult for light from the star to escape, and the light appears dimmer and redder to an observer at a distance.

Eventually, when the star has shrunk to a certain critical radius, the gravitational field at the surface becomes so strong that the light cones are bent inward so much that the light can no longer escape. According to the theory of relativity, nothing can travel faster than light. Thus, if light cannot escape, neither can anything else. Everything is dragged back by the gravitational field. So one has a set of events, a region of space–time, from which it is not possible to escape to reach a distant observer. This region is what we now call a black hole. Its boundary is called the event horizon. It coincides with the paths of the light rays that just fail to escape from the black hole.

In order to understand what you would see if you were watching a star collapse to form a black hole, one has to remember that in the theory of relativity there is no absolute time. Each observer has his own measure of time. The time for someone on a star will be different from that for someone at a distance, because of the gravitational field of the star. This effect has been measured in an experiment on Earth with clocks at the top and bottom of a water tower. Suppose an intrepid

As the star contracts, the gravitational field at its surface gets stronger and the light cones get bent inward more. This makes it more difficult for light from the star to escape, and the light appears dimmer and redder to an observer at a distance.

Three Ways to Grow a Black Hole

Primordial collapse of a bulge

1. Primordial hydrogen cloud collapses around small 'seed' black hole.

2. Infalling gas feeds the hole with more mass and forms stars.

3. Collapse yields a giant elliptical galaxy. Black hole growth stops.

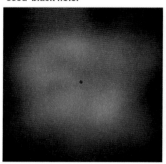
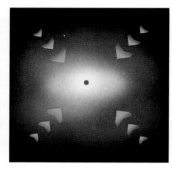
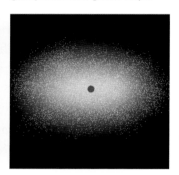

Galaxy collisions

1. Two disk galaxies with central black holes fall toward each other.

2. The galaxies collide, and their cores begin to merge along with their black holes.

3. The merger yields a giant elliptical galaxy with a central black hole that has grown proportionally more massive.

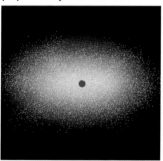

Pseudo bulge

1. Pure disk galaxy forms with, at most, a seed black hole.

2. Disk gas falls into center of galaxy and grows a pseudo bulge which looks like a primordial bulge but really is part of the disk.

3. As pseudo bulge grows, a black hole is created and its mass increases with that of the pseudo bulge.

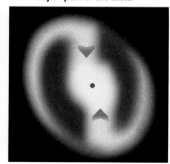
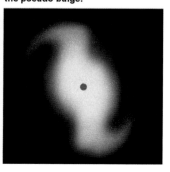

astronaut on the surface of the collapsing star sent a signal every second, according to his watch, to his spaceship orbiting about the star. At some time on his watch, say eleven o'clock, the star would shrink below the critical radius at which the gravitational field became so strong that the signals would no longer reach the spaceship.

His companions watching from the spaceship would find the intervals between successive signals from the astronaut getting longer and longer as eleven o'clock approached. However, the effect would be very small before 10:59:59. They would have to wait only very slightly more than a second between the astronaut's 10:59:58 signal and the one that he sent when his watch read 10:59:59, but they would have to wait forever for the eleven o'clock signal. The light waves emitted from the surface of the star between 10:59:59 and eleven o'clock, by the astronaut's watch, would be spread out over an infinite period of time, as seen from the spaceship.

According to general relativity, there must be a singularity of infinite density within the black hole.

The time interval between the arrival of successive waves at the spaceship would get longer and longer, and so the light from the star would appear redder and redder and fainter and fainter. Eventually the star would be so dim that it could no longer be seen from the spaceship. All that would be left would be a black hole in space. The star would, however, continue to exert the same gravitational force on the spaceship. This is because the star is still visible to the spaceship, at least in principle. It is just that the light from the surface is so red-shifted by the gravitational field of the star that it cannot be seen. However, the red shift does not affect the gravitational field of the star itself. Thus, the spaceship would continue to orbit the black hole.

The work that Roger Penrose and I did between 1965 and 1970 showed that, according to general relativity, there must be a singularity of infinite density within the black hole. This is rather like the big bang at the beginning of time, only it would be an end of time for the collapsing body and the astronaut. At the

The strong

version of the

cosmic censorship

hypothesis states

that in a realistic

solution, the

singularities always

lie either entirely

in the future, like

the singularities

of gravitational

collapse, or

entirely in the past,

like the big bang.

singularity, the laws of science and our ability to predict the future would break down. However, any observer who remained outside the black hole would not be affected by this failure of predictability, because neither light nor any other signal can reach them from the singularity.

This remarkable fact led Roger Penrose to propose the cosmic censorship hypothesis, which might be paraphrased as "God abhors a naked singularity." In other words, the singularities produced by gravitational collapse occur only in places like black holes, where they are decently hidden from outside view by an event horizon. Strictly, this is what is known as the weak cosmic censorship hypothesis: protect observers who remain outside the black hole from the consequences of the breakdown of predictability that occurs at the singularity. But it does nothing at all for the poor unfortunate astronaut who falls into the hole. Shouldn't God protect his modesty as well?

There are some solutions of the equations of general relativity in which it is possible for our astronaut to see a naked singularity. He may be able to avoid hitting the singularity and instead fall through a "worm hole" and come out in another region of the universe. This would offer great possibilities for travel in space and time, but unfortunately it seems that the solutions may all be highly unstable. The least disturbance, such as the presence of an astronaut, may change them so that the astronaut cannot see the singularity until he hits it, and his time comes to an end. In other words, the singularity always lies in his future and never in his past.

The strong version of the cosmic censorship hypothesis states that in a realistic solution, the singularities always lie either entirely in the future, like the singularities of gravitational collapse, or entirely in the past, like the big bang. It is greatly to be hoped that some version of the censorship hypothesis holds, because close to naked singularities it may be possible to travel into the past. While this would be fine for writers of science fiction, it would mean that no one's life would ever be safe.

Someone might go into the past and kill your father or mother before you were conceived.

In a gravitational collapse to form a black hole, the movements would be dammed by the emission of gravitational waves. One would therefore expect that it would not be too long before the black hole would settle down to a stationary state. It was generally supposed that this final stationary state would depend on the details of the body that had collapsed to form the black hole. The black hole might have any shape or size, and its shape might not even be fixed, but instead be pulsating.

The cosmic censorship hypothesis, which might be paraphrased as "God abhors a naked singularity" proposes that the singularities produced by gravitational collapse occur only in places like black holes, where they are "decently" hidden from outside view by an event horizon. Even an astronaut floating in space may not be able to see a singularity until he hits it, and his time comes to an unexpected end.

However, in 1967, the study of black holes was revolutionized by a paper written in Dublin by Werner Israel. Israel showed that any black hole that is not rotating must be perfectly round or spherical. Its size, moreover, would depend only on its mass. It could, in fact, be described by a particular solution of Einstein's equations that had been known since 1917, when it had been found by Karl Schwarzschild shortly after the discovery of general relativity. At first, Israel's result was interpreted by many people, including Israel himself, as evidence that black holes would form only from the collapse of bodies that were perfectly round or spherical. As no real body would be perfectly spherical, this meant that, in general, gravitational collapse would lead to naked singularities. There was, however, a different interpretation of Israel's result, which was advocated by Roger Penrose and John Wheeler in particular. This was that a black hole should behave like a ball of fluid. Although a body might start off in an unspherical state, as it collapsed to form a black hole it would settle down to a spherical state due to the emission of gravitational waves. Further calculations supported this view, and it came to be adopted generally.

Israel's result had dealt only with the case of black holes formed from nonrotating bodies. On the analogy with a ball of fluid, one would expect that a black hole made by the collapse of a rotating body would not be perfectly round. It would have a bulge round the equator caused by the effect of the rotation. We observe a small bulge like this in the sun, caused by its rotation once every twenty-five days or so. In 1963, Roy Kerr, a New Zealander, had found a set of black–hole solutions of the equations of general relativity more general than the Schwarzschild solutions. These "Kerr" black holes rotate at a constant rate, their size and shape depending only on their mass and rate of rotation. If the rotation was zero, the black hole was perfectly round and the solution was identical to the Schwarzschild solution. But if the rotation was nonzero, the black

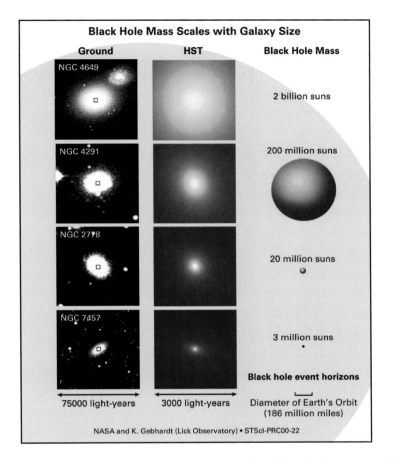

Black Hole Mass Scales with Galaxy Size

Ground	HST	Black Hole Mass
NGC 4649		2 billion suns
NGC 4291		200 million suns
NGC 2778		20 million suns
NGC 7457		3 million suns

Black hole event horizons

| 75000 light-years | 3000 light-years | Diameter of Earth's Orbit (186 million miles) |

NASA and K. Gebhardt (Lick Observatory) • STScI-PRC00-22

This comparison of the hearts of four elliptical galaxies shows that the more massive a galaxy's central bulge of stars, the heftier its black hole. The column of black-and-white pictures at left, taken by ground-based telescopes, shows the galaxies. The inset boxes define the central regions of stars. Close-up images of these regions, as seen by Hubble's Wide Field and Planetary Camera 2, are in the middle column. The column at right lists the masses of the black holes and illustrates the respective diameters of the event horizons. Astronomers determined the mass of each black hole by measuring the motion of stars swirling around it: The closer the stars approach the black hole, the faster their velocity. Astronomers discovered a remarkable new correlation between a black hole's mass and the average speed of the stars in a galaxy's central bulge. The faster the stars are moving, the more massive the black hole. This information suggests that titanic black holes did not precede a host galaxy's birth, but instead co-evolved with the galaxy by trapping a surprisingly exact percentage of the mass of the central hub of stars and gas in a galaxy.

hole bulged outward near its equator. It was therefore natural to conjecture that a rotating body collapsing to form a black hole would end up in a state described by the Kerr solution.

In 1970, a colleague and fellow research student of mine, Brandon Carter, took the first step toward proving this conjecture. He showed that, provided a stationary rotating black hole had an axis of symmetry, like a spinning top, its size and shape would depend only on its mass and rate of rotation. Then, in 1971, I proved that any stationary rotating black hole would indeed have such an axis of symmetry. Finally, in 1973, David Robinson at Kings College, London, used Carter's and my results to show that the conjecture had been correct: Such a black hole had indeed to be the Kerr solution.

So after gravitational collapse a black hole must settle down into a state in which it could be rotating, but not pulsating. Moreover, its size and shape would depend only on its mass and rate of rotation, and not on the nature of the body that had collapsed to form it. This result became known by the maxim "A black hole has no hair." It means that a very large amount of information about the body that has collapsed must be lost when a black hole is formed, because afterward all we can possibly measure about the body is its mass and rate of rotation. The significance of this will be seen in the next lecture. The no-hair theorem is also of great practical importance because it so greatly restricts the possible types of black holes. One can therefore make detailed models of objects that might contain black holes, and compare the predictions of the models with observations.

Black holes are one of only a fairly small number of cases in the history of science where a theory was developed in great detail as a mathematical model before there was any evidence from observations that it was correct. Indeed, this used to be the main argument of opponents of black holes. How could one believe in objects for which the only evidence was calculations based on the dubious theory of general relativity?

The maxim "A black hole has no hair" means that a very large amount of information about the body that has collapsed must be lost when a black hole is formed.

44

In 1963, however, Maarten Schmidt, an astronomer at the Mount Palomar Observatory in California, found a faint, starlike object in the direction of the source of radio waves called 3C273—that is, source number 273 in the third Cambridge catalog of radio sources. When he measured the red shift of the object, he found it was too large to be caused by a gravitational field: If it had been a gravitational red shift, the object would have to be so massive and so near to us that it would disturb the orbits of planets in the solar system. This suggested that the red shift was instead caused by the expansion of the universe, which in turn meant that the object was a very long way away. And to be visible at such a great distance, the object must be very bright and must be emitting a huge amount of energy.

The only mechanism people could think of that would produce such large quantities of energy seemed to be the gravitational collapse not just of a star but of the whole central region of a galaxy. A number of other similar "quasi-stellar objects," or quasars, have since been discovered, all with large red shifts. But they are all too far away, and too difficult, to observe to provide conclusive evidence of black holes.

Further encouragement for the existence of black holes came in 1967 with the discovery by a research student at Cambridge, Jocelyn Bell, of some objects in the sky that were emitting regular pulses of radio waves. At first, Jocelyn and her supervisor, Anthony Hewish, thought that maybe they had made contact with an alien civilization in the galaxy. Indeed, at the seminar at which they announced their discovery, I remember that they called the first four sources to be found LGM 1–4, LGM standing for "Little Green Men."

In the end, however, they and everyone else came to the less romantic conclusion that these objects, which were given the name pulsars, were in fact just rotating neutron stars. They were emitting pulses of radio waves because of a complicated indirection between their magnetic fields and surrounding

A number of other similar "quasi-stellar objects," or quasars, have since been discovered, all with large red shifts. But they are all too far away, and too difficult, to observe to provide conclusive evidence of black holes.

matter. This was bad news for writers of space westerns, but very hopeful for the small number of us who believed in black holes at that time. It was the first positive evidence that neutron stars existed. A neutron star has a radius of about ten miles, only a few times the critical radius at which a star becomes a black hole. If a star could collapse to such a small size, it was not unreasonable to expect that other stars could collapse to even smaller size and become black holes.

How could we hope to detect a black hole, as by its very definition it does not emit any light?

How could we hope to detect a black hole, as by its very definition it does not emit any light? It might seem a bit like looking for a black cat in a coal cellar. Fortunately, there is a way, since as John Michell pointed out in his pioneering paper in 1783, a black hole still exerts a gravitational force on nearby objects. Astronomers have observed a number of systems in which two stars orbit around each other, attracted toward each other by gravity. They also observed systems in which there is only one visible star that is orbiting around some unseen companion.

One cannot, of course, immediately conclude that the companion is a black hole. It might merely be a star that is too faint to be seen. However, some of these systems, like the one called Cygnus X-I, are also strong sources of X-rays. The best explanation for this phenomenon is that the X-rays are generated by matter that has been blown off the surface of the visible star. As it falls toward the unseen companion, it develops a spiral motion—rather like water running out of a bath—and it gets very hot, emitting X-rays. For this mechanism to work, the unseen object has to be very small, like a white dwarf, neutron star, or black hole.

Now, from the observed motion of the visible star, one can determine the lowest possible mass of the unseen object. In the case of Cygnus X-I, this is about six times the mass of the sun. According to Chandrasekhar's result, this is too much for the unseen object to be a white dwarf. It is also too large a mass to be a neutron star. It seems, therefore, that it must be a black hole.

There are other models to explain Cygnus X–I that do not include a black hole, but they are all rather far-fetched. A black hole seems to be the only really natural explanation of the observations. Despite this, I have a bet with Kip Thorne of the California Institute of Technology that in fact Cygnus X–I does not contain a black hole. This is a form of insurance policy for me. I have done a lot of work on black holes, and it would all be wasted if it turned out that black holes do not exist. But in that case, I would have the consolation of winning my bet, which would bring me four years of the magazine *Private Eye.* If black holes do exist, Kip will get only one year of *Penthouse,* because when we made the bet in 1975, we were 80 percent certain that Cygnus was a black hole. By now I would say that we are about 95 percent certain, but the bet has yet to be settled.

There is evidence for black holes in a number of other systems in our galaxy, and for much larger black holes at the centers of other galaxies and quasars. One can also consider the possibility that there might be black holes with masses much less than that of the sun. Such black holes could not be formed by gravitational collapse, because their masses are below the Chandrasekhar mass limit. Stars of this low mass can support themselves against the force of gravity even when they have exhausted their nuclear fuel. So, low-mass black holes could form only if matter were compressed to enormous densities by very large external pressures. Such conditions could occur in a very big hydrogen bomb. The physicist John Wheeler once calculated that if one took all the heavy water in all the oceans of the world, one could build a hydrogen bomb that would compress matter at the center so much that a black hole would be created. Unfortunately, however, there would be no one left to observe it.

A more practical possibility is that such low–mass black holes might have been formed in the high temperatures and pressures

There is evidence for black holes in a number of other systems in our galaxy, and for much larger black holes at the centers of other galaxies and quasars.

of the very early universe. Black holes could have been formed if the early universe had not been perfectly smooth and uniform, because then a small region that was denser than average could be compressed in this way to form a black hole. But we know that there must have been some irregularities, because otherwise the matter in the universe would still be perfectly uniformly distributed at the present epoch, instead of being clumped together in stars and galaxies.

Whether or not the irregularities required to account for stars and galaxies would have led to the formation of a significant number of these primordial black holes depends on the details of the conditions in the early universe. So if we could determine how many primordial black holes there are now, we would learn a lot about the very early stages of the universe. Primordial black holes with masses more than a thousand million tons— the mass of a large mountain—could be detected only by their gravitational influence on other visible matter or on the expansion of the universe. However, as we shall learn in the next lecture, black holes are not really black after all: They glow like a hot body, and the smaller they are, the more they glow. So, paradoxically, smaller black holes might actually turn out to be easier to detect than large ones.

FOURTH LECTURE

BLACK HOLES AIN'T SO BLACK

Before 1970, my research on general relativity had concentrated mainly on the question of whether there had been a big bang singularity. However, one evening in November of that year, shortly after the birth of my daughter, Lucy, I started to think about black holes as I was getting into bed. My disability made this rather a slow process, so I had plenty of time. At that date there was no precise definition of which points in space-time lay inside a black hole and which lay outside.

I had already discussed with Roger Penrose the idea of defining a black hole as the set of events from which it was not possible to escape to a large distance. This is now the generally accepted definition. It means that the boundary of the black hole, the event horizon, is formed by rays of light that just fail to get away from the black hole. Instead, they stay forever, hovering on the edge of the black hole. It is like running away from the police and managing to keep one step ahead but not being able to get clear away.

Suddenly I realized that the paths of these light rays could not be approaching one another, because if they were, they must eventually run into each other. It would be like someone else running away from the police in the opposite direction. You would both be caught or, in this case, fall into a black hole. But if these light rays were swallowed up by the black hole, then they could not have been on the boundary of the black hole. So light rays in the event horizon had to be moving parallel to, or away from, each other.

Another way of seeing this is that the event horizon, the boundary of the black hole, is like the edge of a shadow. It is the

The event horizon is formed by rays of light that just fail to get away from the black hole. Instead, they stay forever, hovering on the edge of the black hole.

The boundaries

of the black hole

according to the

two definitions

would be the

same, provided

the black hole had

settled down to a

stationary state.

edge of the light of escape to a great distance, but, equally, it is the edge of the shadow of impending doom. And if you look at the shadow cast by a source at a great distance, such as the sun, you will see that the rays of light on the edge are not approaching each other. If the rays of light that form the event horizon, the boundary of the black hole, can never approach each other, the area of the event horizon could stay the same or increase with time. It could never decrease, because that would mean that at least some of the rays of light in the boundary would have to be approaching each other. In fact, the area would increase whenever matter or radiation fell into the black hole.

Also, suppose two black holes collided and merged together to form a single black hole. Then the area of the event horizon of the final black hole would be greater than the sum of the areas of the event horizons of the original black holes. This nondecreasing property of the event horizon's area placed an important restriction on the possible behavior of black holes. I was so excited with my discovery that I did not get much sleep that night.

The next day I rang up Roger Penrose. He agreed with me. I think, in fact, that he had been aware of this property of the area. However, he had been using a slightly different definition of a black hole. He had not realized that the boundaries of the black hole according to the two definitions would be the same, provided the black hole had settled down to a stationary state.

THE SECOND LAW OF THERMODYNAMICS

The nondecreasing behavior of a black hole's area was very reminiscent of the behavior of a physical quantity called entropy, which measures the degree of disorder of a system. It is a matter of common experience that disorder will tend to increase if things are left to themselves; one has only to leave a house without repairs to see that. One can create order out of disorder—for example, one can paint the house. However, that requires expenditure of energy, and so decreases the amount of ordered energy available.

A precise statement of this idea is known as the second law of thermodynamics. It states that the entropy of an isolated system never decreases with time. Moreover, when two systems are joined together, the entropy of the combined system is greater than the sum of the entropies of the individual systems. For example, consider a system of gas molecules in a box. The molecules can be thought of as little billiard balls continually colliding with each other and bouncing off the walls of the box. Suppose that initially the molecules are all confined to the left-hand side of the box by a partition. If the partition is then removed, the molecules will tend to spread out and occupy both halves of the box. At some later time they could, by chance, all be in the right half or all be back in the left half. However, it is overwhelmingly more probable that there will be roughly equal numbers in the two halves. Such a state is less ordered, or more disordered, than the original state in which all the molecules were in one half. One therefore says that the entropy of the gas has gone up.

Similarly, suppose one starts with two boxes, one containing oxygen molecules and the other containing nitrogen molecules. If one joins the boxes together and removes the intervening wall, the oxygen and the nitrogen molecules will start to mix. At a later time, the most probable state would be to have a thoroughly uniform mixture of oxygen and nitrogen molecules throughout the two boxes. This state would be less ordered, and hence have more entropy, than the initial state of two separate boxes.

The second law of thermodynamics has a rather different status than that of other laws of science. Other laws, such as Newton's law of gravity, for example, are absolute law—that is, they always hold. On the other hand, the second law is a statistical law—that is, it does not hold always, just in the vast

The molecules can be thought of as little billiard balls continually colliding with each other and bouncing off the walls of the box.

majority of cases. The probability of all the gas molecules in our box being found in one half of the box at a later time is many millions of millions to one, but it could happen.

However, if one has a black hole around, there seems to be a rather easier way of violating the second law: Just throw some matter with a lot of entropy, such as a box of gas, down the black hole. The total entropy of matter outside the black hole would go down. One could, of course, still say that the total entropy, including the entropy inside the black hole, has not gone down. But since there is no way to look inside the black hole, we cannot see how much entropy the matter inside it has. It would be nice, therefore, if there was some feature of the black hole by which observers outside the black hole could tell its entropy; this should increase whenever matter carrying entropy fell into the black hole.

Following my discovery that the area of the event horizon increased whenever matter fell into a black hole, a research student at Princeton named Jacob Bekenstein suggested that the area of the event horizon was a measure of the entropy of the black hole. As matter carrying entropy fell into the black hole, the area of the event horizon would go up, so that the sum of the entropy of matter outside black holes and the area of the horizons would never go down.

This suggestion seemed to prevent the second law of thermodynamics from being violated in most situations. However, there was one fatal flaw: If a black hole has entropy, then it ought also to have a temperature. But a body with a nonzero temperature must emit radiation at a certain rate. It is a matter of common experience that if one heats up a poker in the fire, it glows red hot and emits radiation. However, bodies at lower temperatures emit radiation, too; one just does not normally notice it because the amount is fairly small. This radiation is required in order to prevent violations of the second law. So black holes ought to emit radiation, but by their very definition, black holes are objects that are not supposed to emit

As matter carrying entropy fell into the black hole, the area of the event horizon would go up, so that the sum of the entropy of matter outside black holes and the area of the horizons would never go down.

anything. It therefore seemed that the area of the event horizon of a black hole could not be regarded as its entropy.

In fact, in 1972 I wrote a paper on this subject with Brandon Carter and an American colleague, Jim Bardeen. We pointed

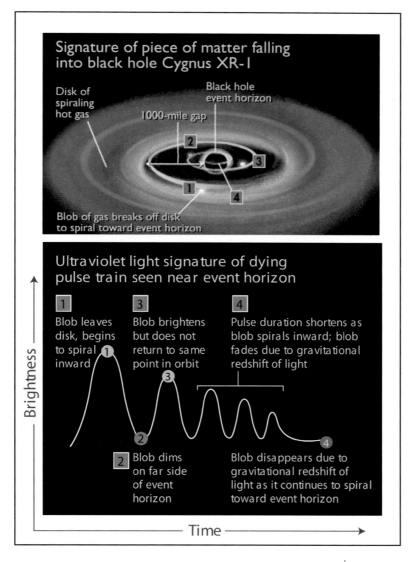

Astronomers may have found evidence for the existence of black holes by observing how matter (like a hot clump of gas) disappears when it falls behind the "event horizon" of a massive, compact object called Cygnus XR-1. When matter falls into a black hole, the area of the event horizon increases.

According to the quantum mechanical uncertainty principle, rotating black holes should create and emit particles.

out that, although there were many similarities between entropy and the area of the event horizon, there was this apparently fatal difficulty. I must admit that in writing this paper I was motivated partly by irritation with Bekenstein, because I felt he had misused my discovery of the increase of the area of the event horizon. However, it turned out in the end that he was basically correct, though in a manner he had certainly not expected.

BLACK HOLE RADIATION

In September 1973, while I was visiting Moscow, I discussed black holes with two leading Soviet experts, Yakov Zeldovich and Alexander Starobinsky. They convinced me that, according to the quantum mechanical uncertainty principle, rotating black holes should create and emit particles. I believed their arguments on physical grounds, but I did not like the mathematical way in which they calculated the emission. I therefore set about devising a better mathematical treatment, which I described at an informal seminar in Oxford at the end of November 1973. At that time I had not done the calculations to find out how much would actually be emitted. I was expecting to discover just the radiation that Zeldovich and Starobinsky had predicted from rotating black holes. However, when I did the calculation, I found, to my surprise and annoyance, that even nonrotating black holes should apparently create and emit particles at a steady rate.

At first I thought that this emission indicated that one of the approximations I had used was not valid. I was afraid if Bekenstein found out about it, he would use it as a further argument to support his ideas about the entropy of black holes, which I still did not like. However, the more I thought about it, the more it seemed that the approximations really ought to hold. But what finally convinced me that the emission was real was that the spectrum of the emitted particles was exactly that which would be emitted by a hot body.

The black hole was emitting particles at exactly the correct rate to prevent violations of the second law.

Since then, the calculations have been repeated in a number of different forms by other people. They all confirm that a black hole ought to emit particles and radiation as if it were a hot body with a temperature that depends only on the black hole's mass: the higher the mass, the lower the temperature. One can understand this emission in the following way: What we think of as empty space cannot be completely empty because that would mean that all the fields, such as the gravitational field and the electromagnetic field, would have to be exactly zero. However, the value of a field and its rate of change with time are like the position and velocity of a particle. The uncertainty principle implies that the more accurately one knows one of these quantities, the less accurately one can know the other.

So in empty space the field cannot be fixed at exactly zero, because then it would have both a precise value, zero, and a precise rate of change, also zero. Instead, there must be a certain minimum amount of uncertainty, or quantum fluctuations, in the value of a field. One can think of these fluctuations as pairs of particles of light or gravity that appear together at some time, move apart, and then come together again and annihilate each other. These particles are called virtual particles. Unlike real particles, they cannot be observed directly with a particle detector. However, their indirect effects, such as small changes in the energy of electron orbits and atoms, can be measured and agree with the theoretical predictions to a remarkable degree of accuracy.

By conservation of energy, one of the partners in a virtual particle pair will have positive energy and the other partner will have negative energy.

By conservation of energy, one of the partners in a virtual particle pair will have positive energy and the other partner will have negative energy. The one with negative energy is condemned to be a short-lived virtual particle. This is because real particles always have positive energy in normal situations. It must therefore seek out its partner and annihilate it. However, the gravitational field inside a black hole is so strong that even a real particle can have negative energy there.

The smaller the

black hole, the less

far the particle

with negative

energy will have

to go before it

becomes a real

particle.

It is therefore possible, if a black hole is present, for the virtual particle with negative energy to fall into the black hole and become a real particle. In this case it no longer has to annihilate its partner; its forsaken partner may fall into the black hole as well. But because it has positive energy, it is also possible for it to escape to infinity as a real particle. To an observer at a distance, it will appear to have been emitted from the black hole. The smaller the black hole, the less far the particle with negative energy will have to go before it becomes a real particle. Thus, the rate of emission will be greater, and the apparent temperature of the black hole will be higher.

The positive energy of the outgoing radiation would be balanced by a flow of negative energy particles into the black hole. By Einstein's famous equation $E = mc2$, energy is equivalent to mass. A flow of negative energy into the black hole therefore reduces its mass. As the black hole loses mass, the area of its event horizon gets smaller, but this decrease in the entropy of the black hole is more than compensated for by the entropy of the emitted radiation, so the second law is never violated.

BLACK HOLE EXPLOSIONS

The lower the mass of the black hole, the higher its temperature is. So as the black hole loses mass, its temperature and rate of emission increase. It therefore loses mass more quickly. What happens when the mass of the black hole eventually becomes extremely small is not quite clear. The most reasonable guess is that it would disappear completely in a tremendous final burst of emission, equivalent to the explosion of millions of H-bombs.

A black hole with a mass a few times that of the sun would have a temperature of only one ten-millionth of a degree above absolute zero. This is much less than the temperature of the microwave radiation that fills the universe, about 2.7 degrees above absolute zero—so such black holes would give off less than they absorb, though even that would be very little. If the universe is destined to go on expanding forever, the temperature of the microwave radiation will eventually decrease to less than that of such a black hole. The hole will then absorb less than it emits and will begin to lose mass. But, even then, its temperature is so low that it would take about 1066 years to evaporate completely. This is much longer than the age of the universe, which is only about 1010 years.

On the other hand, as we learned in the last lecture, there might be primordial black holes with a very much smaller mass that were made by the collapse of irregularities in the very early stages of the universe. Such black holes would have a much higher temperature and would be emitting radiation at a much greater rate. A primordial black hole with an initial mass of a thousand million tons would have a lifetime roughly equal to the age of the universe. Primordial black holes with initial masses less than this figure would already have completely evaporated. However, those with slightly greater masses would still be emitting radiation in the form of X-rays and gamma rays. These are like waves of light, but with a much shorter wavelength. Such holes hardly deserve the epithet black. They really are white hot, and are emitting energy at the rate of about ten thousand megawatts.

One such black hole could run ten large power stations, if only we could harness its output. This would be rather difficult, however. The black hole would have the mass of a mountain compressed into the size of the nucleus of an atom. If you had one of these black holes on the surface of the Earth, there would be no way to stop it falling through the floor to the center of the Earth. It would oscillate through the Earth and back, until eventually it settled down at the center. So the only place to put such a black hole, in which one might use the energy that it emitted, would be in orbit around the Earth. And the only way that one could get it to orbit the Earth would be to attract it there by towing a large mass in front of it, rather like a carrot

There might be primordial black holes with a very much smaller mass that were made by the collapse of irregularities in the very early stages of the universe.

in front of a donkey. This does not sound like a very practical proposition, at least not in the immediate future.

THE SEARCH FOR PRIMORDIAL BLACK HOLES

But even if we cannot harness the emission from these primordial black holes, what are our chances of observing them? We could look for the gamma rays that the primordial black holes emit during most of their lifetime. Although the radiation from most would be very weak because they are far away, the total from all of them might be detectable. We do, indeed, observe such a background of gamma rays. However, this background was probably generated by processes other than primordial black holes. One can say that the observations of the gamma ray background do not provide any positive evidence for primordial black holes. But they tell us that, on average, there cannot be more than three hundred little black holes in every cubic light-year in the universe. This limit means that primordial black holes could make up at most one millionth of the average mass density in the universe.

With primordial black holes being so scarce, it might seem unlikely that there would be one that was near enough for us to observe on its own. But since gravity would draw primordial black holes toward any matter, they should be much more common in galaxies. If they were, say, a million times more common in galaxies, then the nearest black hole to us would probably be at a distance of about a thousand million kilometers, or about as far as Pluto, the farthest known planet. At this distance it would still be very difficult to detect the steady emission of a black hole even if it was ten thousand megawatts.

In order to observe a primordial black hole, one would have to detect several gamma ray quanta coming from the same direction within a reasonable space of time, such as a week.

One can say that the observations of the gamma ray background do not provide any positive evidence for primordial black holes.

Here is an artist's concept of the cosmic fireworks in a spinning, massive black hole. The hole is fueled by a continual in-fall of nearby gas and stars. The gravitational accretion process is far more efficient at converting mass to energy than thermonuclear fusion processes which power individual stars. The extraordinary high pressure and temperature generated near the hole would cause some of the in-falling gas to be ejected along the direction of the black hole's spinning axis to create the galactic jet.

Otherwise, they might simply be part of the background. But Planck's quantum principle tells us that each gamma ray quantum has a very high energy, because gamma rays have a very high frequency. So to radiate even ten thousand megawatts would not take many quanta. And to observe these few quanta coming from the distance of Pluto would require a larger gamma ray detector than any that have been constructed so far. Moreover, the detector would have to be in space, because gamma rays cannot penetrate the atmosphere.

If the black hole

has been emitting

for the last ten or

twenty thousand

million years,

the chances of it

reaching the end of

its life within the

next few years are

really rather small.

Of course, if a black hole as close as Pluto were to reach the end of its life and blow up, it would be easy to detect the final burst of emission. But if the black hole has been emitting for the last ten or twenty thousand million years, the chances of it reaching the end of its life within the next few years are really rather small. It might equally well be a few million years in the past or future. So in order to have a reasonable chance of seeing an explosion before your research grant ran out, you would have to find a way to detect any explosions within a distance of about one light-year. You would still have the problem of needing a large gamma ray detector to observe several gamma ray quanta from the explosion. However, in this case, it would not be necessary to determine that all the quanta came from the same direction. It would be enough to observe that they all arrived within a very short time interval to be reasonably confident that they were coming from the same burst.

One gamma ray detector that might be capable of spotting primordial black holes is the entire Earth's atmosphere. (We are, in any case, unlikely to be able to build a larger detector.) When a high-energy gamma ray quantum hits the atoms in our atmosphere, it creates pairs of electrons and positrons. When these hit other atoms, they in turn create more pairs of electrons and positrons. So one gets what is called an electron shower. The result is a form of light called Cerenkov radiation. One can therefore detect gamma ray bursts by looking for flashes of light in the night sky.

Of course, there are a number of other phenomena, such as lightning, which can also give flashes in the sky. However, one could distinguish gamma ray bursts from such effects by observing flashes simultaneously at two or more thoroughly widely separated locations. A search like this has been carried out by two scientists from Dublin, Neil Porter and Trevor Weekes, using telescopes in Arizona. They found a number of flashes but none that could be definitely ascribed to gamma ray bursts from primordial black holes.

Even if the search for primordial black holes proves negative, as it seems it may, it will still give us important information about the very early stages of the universe. If the early universe had been chaotic or irregular, or if the pressure of matter had been low, one would have expected it to produce many more primordial black holes than the limit set by our observations of the gamma ray background. It is only if the early universe was very smooth and uniform, and with a high pressure, that one can explain the absence of observable numbers of primordial black holes.

GENERAL RELATIVITY AND QUANTUM MECHANICS

Radiation from black holes was the first example of a prediction that depended on both of the great theories of this century, general relativity and quantum mechanics. It aroused a lot of opposition initially because it upset the existing viewpoint: "How can a black hole emit anything?" When I first announced the results of my calculations at a conference at the Rutherford Laboratory near Oxford, I was greeted with general incredulity. At the end of my talk the chairman of the session, John G. Taylor from Kings College, London, claimed it was all nonsense. He even wrote a paper to that effect.

However, in the end most people, including John Taylor, have come to the conclusion that black holes must radiate like hot bodies if our other ideas about general relativity and quantum mechanics are correct. Thus even though we have not yet managed to find a primordial black hole, there is fairly general agreement that if we did, it would have to be emitting a lot of gamma and X-rays. If we do find one, I will get the Nobel Prize.

The existence of radiation from black holes seems to imply that gravitational collapse is not as final and irreversible as we once thought. If an astronaut falls into a black hole, its mass will increase. Eventually, the energy equivalent of that extra mass will

Even if the search for primordial black holes proves negative, as it seems it may, it will still give us important information about the very early stages of the universe.

If an astronaut falls into a black hole, its mass will increase. Eventually, the energy equivalent of that extra mass will be returned to the universe in the form of radiation.

be returned to the universe in the form of radiation. Thus, in a sense, the astronaut will be recycled. It would be a poor sort of immortality, however, because any personal concept of time for the astronaut would almost certainly come to an end as he was crushed out of existence inside the black hole. Even the types of particle that were eventually emitted by the black hole would in general be different from those that made up the astronaut. The only feature of the astronaut that would survive would be his mass or energy.

The approximations I used to derive the emission from black holes should work well when the black hole has a mass greater than a fraction of a gram. However, they will break down at the end of the black hole's life, when its mass gets very small. The most likely outcome seems to be that the black hole would just disappear, at least from our region of the universe. It would take with it the astronaut and any singularity there might be inside the black hole. This was the first indication that quantum mechanics might remove the singularities that were predicted by classical general relativity. However, the methods that I and other people were using in 1974 to study the quantum effects of gravity were not able to answer questions such as whether singularities would occur in quantum gravity.

From 1975 onward, I therefore started to develop a more powerful approach to quantum gravity based on Feynman's idea of a sum over histories. The answers that this approach suggests for the origin and fate of the universe will be described in the next two lectures. We shall see that quantum mechanics allows the universe to have a beginning that is not a singularity. This means that the laws of physics need not break down at the origin of the universe. The state of the universe and its contents, like ourselves, are completely determined by the laws of physics, up to the limit set by the uncertainty principle. So much for free will.

Any personal concept of time for the astronaut would almost certainly come to an end as he was crushed out of existence inside the black hole.

63

FIFTH LECTURE

THE ORIGIN and FATE of the UNIVERSE

Throughout the 1970s I had been working mainly on black holes. However, in 1981 my interest in questions about the origin of the universe was reawakened when I attended a conference on cosmology in the Vatican. The Catholic church had made a bad mistake with Galileo when it tried to lay down the law on a question of science, declaring that the sun went around the Earth. Now, centuries later, it had decided it would be better to invite a number of experts to advise it on cosmology.

At the end of the conference the participants were granted an audience with the pope. He told us that it was okay to study the evolution of the universe after the big bang, but we should not inquire into the big bang itself because that was the moment of creation and therefore the work of God.

I was glad then that he did not know the subject of the talk I had just given at the conference. I had no desire to share the fate of Galileo; I have a lot of sympathy with Galileo, partly because I was born exactly three hundred years after his death.

THE HOT BIG BANG MODEL

In order to explain what my paper was about, I shall first describe the generally accepted history of the universe, according to what is known as the "hot big bang model." This assumes that the universe is described by a Friedmann model, right back to the big bang. In such models one finds that as the universe expands, the temperature of the matter and radiation in it will go down. Since temperature is simply a measure of the average energy of

He told us that it was okay to study the evolution of the universe after the big bang, but we should not inquire into the big bang itself because that was the moment of creation and therefore the work of God.

65

the particles, this cooling of the universe will have a major effect on the matter in it. At very high temperatures, particles will be moving around so fast that they can escape any attraction toward each other caused by the nuclear or electromagnetic forces. But as they cooled off, one would expect particles that attract each other to start to clump together.

At the big bang itself, the universe had zero size and so must have been infinitely hot. But as the universe expanded, the temperature of the radiation would have decreased. One second after the big bang it would have fallen to about ten thousand million degrees. This is about a thousand times the temperature at the center of the sun, but temperatures as high as this are reached in H-bomb explosions. At this time the universe would have contained mostly photons, electrons, and neutrinos and their antiparticles, together with some protons and neutrons.

As the universe continued to expand and the temperature to drop, the rate at which electrons and the electron pairs were being produced in collisions would have fallen below the rate at which they were being destroyed by annihilation. So most of the electrons and antielectrons would have annihilated each other to produce more photons, leaving behind only a few electrons.

About one hundred seconds after the big bang, the temperature would have fallen to one thousand million degrees, the temperature inside the hottest stars. At this temperature, protons and neutrons would no longer have sufficient energy to escape the attraction of the strong nuclear force. They would start to combine together to produce the nuclei of atoms of deuterium, or heavy hydrogen, which contain one proton and one neutron. The deuterium nuclei would then have combined with more protons and neutrons to make helium nuclei, which contained two protons and two neutrons. There would also be small amounts of a couple of heavier elements, lithium and beryllium.

At the big bang itself, the universe had zero size and so must have been infinitely hot. But as the universe expanded, the temperature of the radiation would have decreased.

One can calculate that in the hot big bang model about a quarter of the protons and neutrons would have been converted into helium nuclei, along with a small amount of heavy hydrogen and other elements. The remaining neutrons would have decayed into protons, which are the nuclei of ordinary hydrogen atoms. These predictions agree very well with what is observed.

The hot big bang model also predicts that we should be able to observe the radiation left over from the hot early stages. However, the temperature would have been reduced to a few degrees above absolute zero by the expansion of the universe. This is the explanation of the microwave background of radiation that was discovered by Penzias and Wilson in 1965. We are therefore thoroughly confident that we have the right picture, at least back to about one second after the big bang. Within only a few hours of the big bang, the production of helium and other elements would have stopped. And after that, for the next million years or so, the universe would have just continued expanding, without anything much happening. Eventually, once the temperature had dropped to a few thousand degrees, the electrons and nuclei would no longer have had enough energy to overcome the electromagnetic attraction between them. They would then have started combining to form atoms.

The universe as a whole would have continued expanding and cooling. However, in regions that were slightly denser than average, the expansion would have been slowed down by extra gravitational attraction. This would eventually stop expansion in some regions and cause them to start to recollapse. As they were collapsing, the gravitational pull of matter outside these regions might start them rotating slightly. As the collapsing region got smaller, it would spin faster—just as skaters spinning on ice spin faster as they draw in their arms. Eventually, when the region got small enough, it would be spinning fast enough to balance the attraction of gravity. In this way, disklike rotating galaxies were born.

The hot big bang model also predicts that we should be able to observe the radiation left over from the hot early stages.

As time went on, the gas in the galaxies would break up into smaller clouds that would collapse under their own gravity. As these contracted, the temperature of the gas would increase until it became hot enough to start nuclear reactions. These would

In 1995, the majestic spiral galaxy NGC 4414 was imaged by the Hubble Space Telescope. Based on their discovery and careful brightness measurements of variable stars in this galaxy, the astronomers were able to make an accurate determination of the distance to the galaxy. The resulting distance to NGC 4414, about 60 million light-years, along with similarly determined distances to other nearby galaxies, contributes to astronomers' overall knowledge of the rate of expansion of the universe. In 1999, the Hubble Heritage Team revisited NGC 4414 and created a stunning full-color look at the entire dusty spiral galaxy. The new Hubble picture shows that the central regions of this galaxy, as is typical of most spirals, contain primarily older, yellow and red stars. The outer spiral arms are considerably bluer due to ongoing formation of young, blue stars, the brightest of which can be seen individually at the high resolution provided by the Hubble camera.

convert the hydrogen into more helium, and the heat given off would raise the pressure, and so stop the clouds from contracting any further. They would remain in this state for a long time as stars like our sun, burning hydrogen into helium and radiating the energy as heat and light.

More massive stars would need to be hotter to balance their stronger gravitational attraction. This would make the nuclear fusion reactions proceed so much more rapidly that they would use up their hydrogen in as little as a hundred million years. They would then contract slightly and, as they heated up further, would start to convert helium into heavier elements like carbon or oxygen. This, however, would not release much more energy, so a crisis would occur, as I described in my lecture on black holes.

What happens next is not completely clear, but it seems likely that the central regions of the star would collapse to a very dense state, such as a neutron star or black hole. The outer regions of the star may get blown off in a tremendous explosion called a supernova, which would outshine all the other stars in the galaxy. Some of the heavier elements produced near the end of the star's life would be flung back into the gas in the galaxy. They would provide some of the raw material for the next generation of stars.

Our own sun contains about 2 percent of these heavier elements because it is a second- or third-generation star. It was formed some five thousand million years ago out of a cloud of rotating gas containing the debris of earlier supernovas. Most of the gas in that cloud went to form the sun or got blown away. However, a small amount of the heavier elements collected together to form the bodies that now orbit the sun as planets like the Earth.

The outer regions of the star may get blown off in a tremendous explosion called a supernova, which would outshine all the other stars in the galaxy.

OPEN QUESTIONS

This picture of a universe that started off very hot and cooled as it expanded is in agreement with all the observational evidence

If the rate of

expansion one

second after the

big bang had been

smaller by even

one part in a

hundred thousand

million million,

the universe would

have recollapsed

before it ever

reached its

present size.

that we have today. Nevertheless, it leaves a number of important questions unanswered. First, why was the early universe so hot? Second, why is the universe so uniform on a large scale—why does it look the same at all points of space and in all directions?

Third, why did the universe start out with so nearly the critical rate of expansion to just avoid recollapse? If the rate of expansion one second after the big bang had been smaller by even one part in a hundred thousand million million, the universe would have recollapsed before it ever reached its present size. On the other hand, if the expansion rate at one second had been larger by the same amount, the universe would have expanded so much that it would be effectively empty now.

Fourth, despite the fact that the universe is so uniform and homogenous on a large scale, it contains local lumps such as stars and galaxies. These are thought to have developed from small differences in the density of the early universe from one region to another. What was the origin of these density fluctuations?

The general theory of relativity, on its own, cannot explain these features or answer these questions. This is because it predicts that the universe started off with infinite density at the big bang singularity. At the singularity, general relativity and all other physical laws would break down. One cannot predict what would come out of the singularity. As I explained before, this means that one might as well cut any events before the big bang out of the theory, because they can have no effect on what we observe. Space–time would have a boundary—a beginning at the big bang. Why should the universe have started off at the big bang in just such a way as to lead to the state we observe today? Why is the universe so uniform, and expanding at just the critical rate to avoid recollapse? One would feel happier about this if one could show that quite a number of different initial configurations for the universe would have evolved to produce a universe like the one we observe. If this is the case, a universe that developed from some sort of random initial conditions should contain a number of regions that are like what we observe. There might also be regions that were very different. However, these regions would probably not be suitable for the formation of galaxies and stars. These are essential prerequisites for the development of intelligent life, at least as we know it. Thus, these regions would not contain any beings to observe that they were different.

When one considers cosmology, one has to take into account the selection principle that we live in a region of the universe that is suitable for intelligent life. This fairly obvious and elementary consideration is sometimes called the anthropic principle. Suppose, on the other hand, that the initial state of the universe had to be chosen extremely carefully to lead to something like what we see around us. Then the universe would be unlikely to contain any region in which life would appear.

In the hot big bang model that I described earlier, there was not enough time in the early universe for heat to have flowed from one region to another. This means that different regions of the universe would have had to have started out with exactly the same temperature in order to account for the fact that the microwave background has the same temperature in every direction we look. Also, the initial rate of expansion would have had to be chosen very precisely for the universe not to have recollapsed before now. This means that the initial state of the universe must have been very carefully chosen indeed if the hot big bang model was correct right back to the beginning of time. It would be very difficult to explain why the universe should have begun in just this way, except as the act of a God who intended to create beings like us.

The initial rate of expansion would have had to be chosen very precisely for the universe not to have recollapsed before now.

THE INFLATIONARY MODEL

In order to avoid this difficulty with the very early stages of the hot big bang model, Alan Guth at the Massachusetts Institute of Technology put forward a new model. In this, many different initial configurations could have evolved to something like the present universe. He suggested that the early universe might have had a period of very rapid, or exponential, expansion. This expansion is said to be inflationary—an analogy with the inflation in prices that occurs to a greater or lesser degree in every country. The world record for price inflation was probably in Germany after the first war, when the price of a loaf of bread went from under a mark to millions of marks in a few months. But the inflation we think may have occurred in the size of the universe was much greater even than that—a million million million million million times in only a tiny fraction of a second. Of course, that was before the present government.

Guth suggested that the universe started out from the big bang very hot. One would expect that at such high temperatures, the strong and weak nuclear forces and the electromagnetic force would all be unified into a single force. As the universe expanded, it would cool, and particle energies would go down. Eventually there would be what is called a phase transition, and the symmetry between the forces would be broken. The strong force would become different from the weak and electromagnetic forces. One common example of a phase transition is the freezing of water when you cool it down. Liquid water is symmetrical, the same at every point and in every direction. However, when ice crystals form, they will have definite positions and will be lined up in some direction. This breaks the symmetry of the water.

In the case of water, if one is careful, one can "supercool" it. That is, one can reduce the temperature below the freezing point—0 degrees centigrade—without ice forming. Guth suggested that the universe might behave in a similar way: The temperature might drop below the critical value without the symmetry between the forces being broken. If this happened, the universe would be in an unstable state, with more energy than if the symmetry had been broken. This special extra energy can be shown to have an antigravitational effect. It would act just like a cosmological constant.

Liquid water is symmetrical, the same at every point and in every direction.

Einstein introduced the cosmological constant into general relativity when he was trying to construct a static model of the universe. However, in this case, the universe would already be expanding. The repulsive effect of this cosmological constant would therefore have made the universe expand at an ever-increasing rate. Even in regions where there were more matter particles than average, the gravitational attraction of the matter would have been outweighed by the repulsion of the effective cosmological constant. Thus, these regions would also expand in an accelerating inflationary manner.

As the universe expanded, the matter particles got farther apart. One would be left with an expanding universe that contained hardly any particles. It would still be in the supercooled state, in which the symmetry between the forces is not broken. Any irregularities in the universe would simply have been smoothed out by the expansion, as the wrinkles in a balloon are smoothed away when you blow it up. Thus, the present smooth and uniform state of the universe could have evolved from many different nonuniform initial states. The rate of expansion would also tend toward just the critical rate needed to avoid recollapse.

Any irregularities in the universe would simply have been smoothed out by the expansion, as the wrinkles in a balloon are smoothed away when you blow it up.

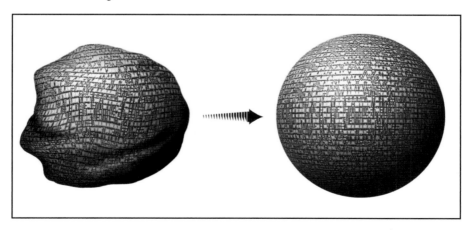

The present smooth and uniform state of the universe could have evolved from many different nonuniform initial states.

When the universe

doubles in size,

the positive

matter energy

and the negative

gravitational

energy both

double, so the

total energy

remains zero.

Moreover, the idea of inflation could also explain why there is so much matter in the universe. There are something like 1,080 particles in the region of the universe that we can observe. Where did they all come from? The answer is that, in quantum theory, particles can be created out of energy in the form of particle/antiparticle pairs. But that just raises the question of where the energy came from. The answer is that the total energy of the universe is exactly zero.

The matter in the universe is made out of positive energy. However, the matter is all attracting itself by gravity. Two pieces of matter that are close to each other have less energy than the same two pieces a long way apart. This is because you have to expend energy to separate them. You have to pull against the gravitational force attracting them together. Thus, in a sense, the gravitational field has negative energy. In the case of the whole universe, one can show that this negative gravitational energy exactly cancels the positive energy of the matter. So the total energy of the universe is zero.

Now, twice zero is also zero. Thus, the universe can double the amount of positive matter energy and also double the negative gravitational energy without violation of the conservation of energy. This does not happen in the normal expansion of the universe in which the matter energy density goes down as the universe gets bigger. It does happen, however, in the inflationary expansion, because the energy density of the supercooled state remains constant while the universe expands. When the universe doubles in size, the positive matter energy and the negative gravitational energy both double, so the total energy remains zero. During the inflationary phase, the universe increases its size by a very large amount. Thus, the total amount of energy available to make particles becomes very large. As Guth has remarked, "It is said that there is no such thing as a free lunch. But the universe is the ultimate free lunch."

THE END OF INFLATION

The universe is not expanding in an inflationary way today. Thus, there had to be some mechanism that would eliminate the very large effective cosmological constant. This would change the rate of expansion from an accelerated one to one that is slowed down by gravity, as we have today. As the universe expanded and cooled, one might expect that eventually the symmetry between the forces would be broken, just as supercooled water always freezes in the end. The extra energy of the unbroken symmetry state would then be released and would reheat the universe. The universe would then go on to expand and cool, just like the hot big bang model. However, there would now be an explanation of why the universe was expanding at exactly the critical rate and why different regions had the same temperature.

In Guth's original proposal, the transition to broken symmetry was supposed to occur suddenly, rather like the appearance of ice crystals in very cold water. The idea was that "bubbles" of the new phase of broken symmetry would have formed in the old phase, like bubbles of steam surrounded by boiling water. The bubbles were supposed to expand and meet up with each other until the whole universe was in the new phase. The trouble was, as I and several other people pointed out, the universe was expanding so fast that the bubbles would be moving away from each other too rapidly to join up. The universe would be left in a very nonuniform state, with some regions having symmetry between the different forces. Such a model of the universe would not correspond to what we see.

In October 1981 I went to Moscow for a conference on quantum gravity. After the conference, I gave a seminar on the inflationary model and its problems at the Sternberg Astronomical Institute. In the audience was a young Russian, Andrei Linde. He said that the difficulty with the bubbles not joining up could be avoided if the bubbles were very big. In this

In Guth's original proposal, the transition to broken symmetry was supposed to occur suddenly, rather like the appearance of ice crystals in very cold water.

75

case, our region of the universe could be contained inside a single bubble. In order for this to work, the change from symmetry to broken symmetry must have taken place very slowly inside the bubble, but this is quite possible according to grand unified theories.

Linde's idea of a slow breaking of symmetry was very good, but I pointed out that his bubbles would have been bigger than the size of the universe at the time. I showed that instead the symmetry would have broken everywhere at the same time, rather than just inside bubbles. This would lead to a uniform universe, like we observe. The slow symmetry breaking model was a good attempt to explain why the universe is the way it is. However, I and several other people showed that it predicted much greater variations in the microwave background radiation than are observed. Also, later work cast doubt on whether there would have been the right kind of phase transition in the early universe. A better model, called the chaotic inflationary model, was introduced by Linde in 1983. This doesn't depend on phase transitions, and it can give us the right size of variations of the microwave background. The inflationary model showed that the present state of the universe could have arisen from quite a large number of different initial configurations. It cannot be the case, however, that every initial configuration would have led to a universe like the one we observe. So even the inflationary model does not tell us why the initial configuration was such as to produce what we observe. Must we turn to the anthropic principle for an explanation? Was it all just a lucky chance? That would seem a counsel of despair, a negation of all our hopes of understanding the underlying order of the universe.

The slow symmetry breaking model was a good attempt to explain why the universe is the way it is.

QUANTUM GRAVITY

In order to predict how the universe should have started off, one needs laws that hold at the beginning of time. If the classical theory of general relativity was correct, the singularity theorem showed that the beginning of time would have been a point of infinite density and curvature. All the known laws of science would break down at such a point. One might suppose that there were new laws that held at singularities, but it would be very difficult even to formulate laws at such badly behaved points and we would have no guide from observations as to what those laws might be. However, what the singularity theorems really indicate is that the gravitational field becomes so strong that quantum gravitational effects become important: Classical theory is no longer a good description of the universe. So one has to use a quantum theory of gravity to discuss the very early stages of the universe. As we shall see, it is possible in the quantum theory for the ordinary laws of science to hold everywhere, including at the beginning of time. It is not necessary to postulate new laws for singularities, because there need not be any singularities in the quantum theory.

We don't yet have a complete and consistent theory that combines quantum mechanics and gravity. However, we are thoroughly certain of some features that such a unified theory should have. One is that it should incorporate Feynman's proposal to formulate quantum theory in terms of a sum over histories. In this approach, a particle going from A to B does not have just a single history as it would in a classical theory. Instead, it is supposed to follow every possible path in space–time. With each of these histories, there are associated a couple of numbers, one representing the size of a wave and the other representing its position in the cycle—its phase.

The probability that the particle, say, passes through some particular point is found by adding up the waves associated with every possible history that passes through that point. When one actually tries to perform these sums, however, one runs into severe technical problems. The only way around these is the following peculiar prescription: One must add up the waves for particle histories that are not in the real time that you and I experience but take place in imaginary time.

We don't yet have a complete and consistent theory that combines quantum mechanics and gravity. However, we are thoroughly certain of some features that such a unified theory should have.

Imaginary time may sound like science fiction, but it is in fact a well–defined mathematical concept. To avoid the technical difficulties with Feynman's sum over histories, one must use imaginary time. This has an interesting effect on space–time: The distinction between time and space disappears completely. A space–time in which events have imaginary values of the time coordinate is said to be Euclidean because the metric is positive definite.

In Euclidean space-time there is no difference between the time direction and directions in space. On the other hand, in real space-time, in which events are labeled by real values of the time coordinate, it is easy to tell the difference. The time direction lies within the light cone, and space directions lie outside. One can regard the use of imaginary time as merely a mathematical device—or trick—to calculate answers about real space-time. However, there may be more to it than that. It may be that

In Euclidean space-time there is no difference between the time direction and directions in space.

Euclidean space-time is the fundamental concept and what we think of as real space-time is just a figment of our imagination.

When we apply Feynman's sum over histories to the universe, the analogue of the history of a particle is now a complete curved space–time which represents the history of the whole universe. For the technical reasons mentioned above, these curved space–times must be taken to be Euclidean. That is, time is imaginary and is indistinguishable from directions in space. To calculate the probability of finding a real space-time with some certain property, one adds up the waves associated with all the histories in imaginary time that have that property. One can then work out what the probable history of the universe would be in real time.

THE NO BOUNDARY CONDITION

In the classical theory of gravity, which is based on real space-time, there are only two possible ways the universe can behave. Either it has existed for an infinite time, or else it had a beginning at a singularity at some finite time in the past. In fact, the singularity theorems show it must be the second possibility. In the quantum theory of gravity, on the other hand, a third possibility arises. Because one is using Euclidean space-times, in which the time direction is on the same footing as directions in space, it is possible for space-time to be finite in extent and yet to have no singularities that formed a boundary or edge. Space-time would be like the surface of the Earth, only with two more dimensions. The surface of the Earth is finite in extent but it doesn't have a boundary or edge. If you sail off into the sunset, you don't fall off the edge or run into a singularity. I know, because I have been around the world.

If Euclidean space-times direct back to infinite imaginary time or else started at a singularity, we would have the same problem as in the classical theory of specifying the initial state

It is possible for space–time to be finite in extent and yet to have no singularities that formed a boundary or edge.

The astronauts onboard the Space Shuttle Columbia took this 70mm picture of the Earth's surface featuring the Sinai Peninsula and the Dead Sea Rift. In the quantum theory of gravity, space-time would be like the surface of the Earth: finite in extent but without a boundary or edge.

of the universe. God may know how the universe began, but we cannot give any particular reason for thinking it began one way rather than another. On the other hand, the quantum theory of gravity has opened up a new possibility. In this, there would be no boundary to space–time. Thus, there would be no need to specify the behavior at the boundary. There would be no singularities at which the laws of science broke down and no edge of space–time at which one would have to appeal to God or some new law to set the boundary conditions for space-time. One could say: "The boundary condition of the universe is that it has no boundary." The universe would be completely self-contained and not affected by anything outside itself. It would be neither created nor destroyed. It would just be.

It was at the conference in the Vatican that I first put forward the suggestion that maybe time and space together formed a surface that was finite in size but did not have any boundary or edge. My paper was rather mathematical, however, so its implications for the role of God in the creation of the universe were not noticed at the time—just as well for me. At the time of the Vatican conference, I did not know how to use a no boundary idea to make predictions about the universe. However, I spent the following summer at the University of California, Santa Barbara. There, a friend and colleague of mine, Jim Hartle, worked out with me what conditions the universe must satisfy if space–time had no boundary. I should emphasize that this idea that time and space should be finite without boundary is just a proposal. It cannot be deduced from some other principle. Like any other scientific theory, it may initially be put forward for aesthetic or metaphysical reasons, but the real test is whether it makes predictions that agree with observation. This, however, is difficult to determine in the case of quantum gravity, for two reasons. First, we are not yet sure exactly which theory successfully combines general relativity and quantum mechanics, though we know quite a lot about the form such a theory must have. Second, any model that described the whole universe in detail would be much too complicated mathematically for us to be able to calculate exact predictions. One therefore has to make approximations—and even then, the problem of extracting predictions remains a difficult one.

One finds, under the no boundary proposal, that the chance of the universe being found to be following most of the possible histories is negligible. But there is a particular family of histories that are much more probable than the others. These histories may be pictured as being like the surface of the Earth, with a

Any model that described the whole universe in detail would be much too complicated mathematically for us to be able to calculate exact predictions.

The universe expanding and contracting in imaginary time.

distance from the North Pole representing imaginary time; the size of a circle of latitude would represent the spatial size of the universe. The universe starts at the North Pole as a single point. As one moves south, the circle of latitude get bigger, corresponding to the universe expanding with imaginary time. The universe would reach a maximum size at the equator and would contract again to a single point at the South Pole. Even though the universe would have zero size at the North and South poles, these points would not be singularities any more than the North and South poles on the Earth are singular. The laws of science will hold at the beginning of the universe, just as they do at the North and South poles on the Earth.

The history of the universe in real time, however, would look very different. It would appear to start at some minimum size, equal to the maximum size of the history in imaginary time. The universe would then expand in real time like the inflationary model. However, one would not now have to assume that the universe was created somehow in the right sort of state. The universe would expand to a very large size, but eventually it would collapse again into what looks like a singularity in real time. Thus, in a sense, we are still all doomed, even if we keep away from black holes. Only if we could picture the universe in terms of imaginary time would there be no singularities.

The singularity theorems of classical general relativity showed that the universe must have a beginning, and that this beginning must be described in terms of quantum theory. This in turn led to the idea that the universe could be finite in imaginary time, but without boundaries or singularities. When one goes back to the real time in which we live, however, there will still appear to be singularities. The poor astronaut who falls into a black hole will still come to a sticky end. It is only if he could live in imaginary time that he would encounter no singularities. This might suggest that the so–called imaginary time is really the fundamental time, and that what we call real time is something we create just in our minds. In real time, the universe has a beginning and an end at singularities that form a boundary to space-time and at which the laws of science break down. But in imaginary time, there are no singularities or boundaries. So maybe what we call imaginary time is really more basic, and what we call real time is just an idea that we invent to help us describe what we think the universe is like. But according to the approach I described in the first lecture, a scientific theory is just a mathematical model we make to

Maybe what we call imaginary time is really more basic, and what we call real time is just an idea that we invent to help us describe what we think the universe is like.

Artist's View of Star Formation in the Early Universe
Painting by Adolf Schaller • STScI-PRC02-02

The deepest views of the cosmos from the Hubble Space Telescope yield clues that the very first stars may have burst into the universe as brilliantly and spectacularly as a fireworks finale. Except in this case, the finale came first, long before Earth, the sun and the Milky Way galaxy formed. Studies of Hubble's deepest views of the heavens lead to the preliminary conclusion that the universe made a significant portion of its stars in a torrential firestorm of star birth, which abruptly lit up the pitch-dark heavens just a few hundred million years after the "big bang," the tremendous explosion that created the cosmos. Though stars continue to be born today in galaxies, the star birth rate could be a trickle compared to the predicted gusher of stars in those opulent early years.

describe our observations. It exists only in our minds. So it does not have any meaning to ask: Which is real, "real" or "imaginary" time? It is simply a matter of which is a more useful description.

The no boundary proposal seems to predict that, in real time, the universe should behave like the inflationary models. A particularly interesting problem is the size of the small departures from uniform density in the early universe. These are thought to have led to the formation first of the galaxies, then of stars, and finally of beings like us. The uncertainty principle implies that the early universe cannot have been completely uniform. Instead, there must have been some uncertainties or fluctuations in the positions and velocities of the particles. Using the no boundary condition, one finds that the universe must have started off with just the minimum possible nonuniformity allowed by the uncertainty principle.

The universe would have then undergone a period of rapid expansion, like in the inflationary models. During this period, the initial nonuniformities would have been amplified until they could have been big enough to explain the origin of galaxies. Thus, all the complicated structures that we see in the universe might be explained by the no boundary condition for the universe and the uncertainty principle of quantum mechanics.

The idea that space and time may form a closed surface without boundary also has profound implications for the role of God in the affairs of the universe. With the success of scientific theories in describing events, most people have come to believe that God allows the universe to evolve according to a set of laws. He does not seem to intervene in the universe to break these laws. However, the laws do not tell us what the universe should have looked like when it started. It would still be up to God to wind up the clockwork and choose how to start it off. So long as the universe had a beginning that was a singularity, one could suppose that it was created by an outside agency. But if the universe is really completely self-contained, having no boundary or edge, it would be neither created nor destroyed. It would simply be. What place, then, for a creator?

The idea that space and time may form a closed surface without boundary also has profound implications for the role of God in the affairs of the universe.

The DIRECTION of TIME

In his book, *The Go Between,* L. P. Hartley wrote, "The past is a foreign country. They do things differently there—but why is the past so different from the future? Why do we remember the past, but not the future?" In other words, why does time go forward? Is this connected with the fact that the universe is expanding?

C, P, T

The laws of physics do not distinguish between the past and the future. More precisely, the laws of physics are unchanged under the combination of operations known as C, P, and T. (C means changing particles for antiparticles. P means taking the mirror image so left and right are swapped for each other. And T means reversing the direction of motion of all particles—in effect, running the motion backward.) The laws of physics that govern the behavior of matter under all normal situations are unchanged under the operations C and P on their own. In other words, life would be just the same for the inhabitants of another planet who were our mirror images and who were made of antimatter. If you meet someone from another planet, and he holds out his left hand, don't shake it. He might be made of antimatter. You would both disappear in a tremendous flash of light. If the laws of physics are unchanged by the combination of operations C and P, and also by the combination C, P, and T, they must also be unchanged under the operation T alone. Yet, there is a big difference between the forward and backward directions of time in ordinary life. Imagine a cup of water falling off a table and

Life would be just the same for the inhabitants of another planet who were our mirror images and who were made of antimatter.

There is a big difference between the forward and backward directions of time in ordinary life.

breaking in pieces on the floor. If you take a film of this, you can easily tell whether it is being run forward or backward. If you run it backward, you will see the pieces suddenly gather themselves together off the floor and jump back to form a whole cup on the table. You can tell that the film is being run backward because this kind of behavior is never observed in ordinary life. If it were, the crockery manufacturers would go out of business.

THE ARROWS OF TIME

An intact cup on the table is a state of high order, but a broken cup on the floor is a disordered state.

The explanation that is usually given as to why we don't see broken cups jumping back onto the table is that it is forbidden by the second law of thermodynamics. This says that disorder or entropy always increases with time. In other words, it is Murphy's Law—things get worse. An intact cup on the table is a state of high order, but a broken cup on the floor is a disordered state. One can therefore go from the whole cup on the table in the past to the broken cup on the floor in the future, but not the other way around.

The increase of disorder or entropy with time is one example of what is called an *arrow of time,* something that gives a direction to time and distinguishes the past from the future. There are at least three different arrows of time. First, there is the thermodynamic arrow of time—the direction of time in which disorder or entropy increases. Second, there is the psychological arrow of time. This is the direction in which we feel time passes—the direction of time in which we remember the past, but not the future. Third, there is the cosmological arrow of time. This is the direction of time in which the universe is expanding rather than contracting.

I shall argue the the pyschological arrow is determined by the thermodynamic arrow and that these two arrows always point in the same direction. If one makes the no boundary assumption for the universe, they are related to the cosmological arrow of time, though they may not point in the same direction. However, I shall argue that it is only when they agree with the cosmological arrow that there will be intelligent beings who can ask the question: Why does disorder increase in the same direction of time as that in which the universe expands?

The second law of thermodynamics is based on the fact that there are many more disordered states than there are ordered ones.

THE THERMODYNAMIC ARROW

I shall talk first about the thermodynamic arrow of time. The second law of thermodynamics is based on the fact that there are many more disordered states than there are ordered ones. For example, consider the pieces of a jigsaw in a box. There is one, and only one, arrangement in which the pieces make a complete picture. On the other hand, there are a very large number of arrangements in which the pieces are disordered and don't make a picture.

Suppose a system starts out in one of the small number of ordered states. As time goes by, the system will evolve according to the laws of physics and its state will change. At a later time, there is a high probability that it will be in a more disordered state, simply because there are so many more disordered states. Thus, disorder will tend to increase with time if the system obeys an initial condition of high order.

Disorder of the pieces will probably increase with time if they obey the initial condition that they start in a state of high order.

Suppose the pieces of the jigsaw start off in the ordered arrangement in which they form a picture. If you shake the box, the pieces will take up another arrangement. This will probably be a disordered arrangement in which the pieces don't form a proper picture, simply because there are so many more disordered arrangements. Some groups of pieces may still form parts of the picture, but the more you shake the box, the more likely it is that these groups will get broken up. The pieces will take up a completely jumbled state in which they don't form any sort of picture. Thus, the disorder of the pieces will probably increase with time if they obey the initial condition that they start in a state of high order.

Suppose, however, that God decided that the universe should finish up at late times in a state of high order but it didn't matter what state it started in. Then, at early times the universe would probably be in a disordered state, and disorder would decrease with time. You would have broken cups gathering themselves together and jumping back on the table. However, any human beings observing the cups would be living in a universe in which disorder decreased with time. I shall argue that such beings would have a psychological arrow of time that was backward. That is, they would remember thence at late times and not remember thence at early times.

THE PSYCHOLOGICAL ARROW

It is rather difficult to talk about human memory because we don't know how the brain works in detail. We do, however, know all about how computer memories work. I shall therefore discuss the psychological arrow of time for computers. I think it is reasonable to assume that the arrow for computers is the same as that for human. If it were not, one could make a killing on the stock exchange by having a computer that would remember tomorrow's prices.

A computer memory is basically some device that can be in either one of two states. An example would be a superconducting loop of wire. If there is an electric current flowing in the loop, it will continue to flow because there is no resistance. On the other hand, if there is no current, the loop will continue without a current. One can label the two states of the memory "one" and "zero."

Before an item is recorded in the memory, the memory is in a disordered state with equal probabilities for one and zero. After the memory interacts with the system to be remembered, it will definitely be in one state or the other, according to the state of the system. Thus, the memory passes from a disordered state to an ordered one. However, in order to make sure that the memory is in the right state, it is necessary to use a certain amount of energy. This energy is dissipated as heat and increases the amount of disorder in the universe. One can show that this increase of disorder is greater than the increase in the order of the memory. Thus, when a computer records an item in memory, the total amount of disorder in the universe goes up.

The direction of time in which a computer remembers the past is the same as that in which disorder increases. This means that our subjective sense of the direction of time, the psychological arrow of time, is determined by the thermodynamic arrow of time. This makes the second law of thermodynamics almost trivial. Disorder increases with time because we measure time in the direction in which disorder increases. You can't have a safer bet than that.

THE BOUNDARY CONDITIONS OF THE UNIVERSE

But why should the universe be in a state of high order at one end of time, the end that we call the past? Why was it not in a state of complete disorder at all times? After all, this might seem more

Our subjective sense of the direction of time, the psychological arrow of time, is determined by the thermodynamic arrow of time.

probable. And why is the direction of time in which disorder increases the same as that in which the universe expands?

One possible answer is that God simply chose that the universe should be in a smooth and ordered state at the beginning of the expansion phase. We should not try to understand why or question His reasons because the beginning of the universe was the work of God. But the whole history of the universe can be said to be the work of God.

It appears that the universe evolves according to well-defined laws. These laws may or may not be ordained by God, but it seems that we can discover and understand them.

It appears that the universe evolves according to well-defined laws. These laws may or may not be ordained by God, but it seems that we can discover and understand them. Is it, therefore, unreasonable to hope that the same or similar laws may also hold at the beginning of the universe? In the classical theory of general relativity, the beginning of the universe has to be a singularity of infinite density in space-time curvature. Under such conditions, all the known laws of physics would break down. Thus, one could not use them to predict how the universe would begin.

The universe could have started out in a very smooth and ordered state. This would have led to well-defined thermodynamic and cosmological arrows of time, like we observe. But it could equally well have started out in a very lumpy and disordered state. In this case, the universe would already be in a state of complete disorder, so disorder could not increase with time. It would either stay constant, in which case there would be no well-defined thermodynamic arrow of time, or it would decrease, in which case the thermodynamic arrow of time would point in the opposite direction to the cosmological arrow. Neither of these possibilities would agree with what we observe.

As I mentioned, the classical theory of general relativity predicts that the universe should begin with a singularity where the curvature of space-time is infinite. In fact, this means that classical general relativity predicts its own downfall. When the curvature of space-time becomes large, quantum gravitational effects will become important and the classical theory will cease to be a good description of the universe. One has to use the quantum theory of gravity to understand how the universe began.

In a quantum theory of gravity, one considers all possible histories of the universe. Associated with each history, there are a couple of numbers. One represents the size of a wave and the other the face of the wave, that is, whether the wave is at a crest or a trough. The probability of the universe having a particular property is given by adding up the waves for all the histories with that property. The histories would be curved spaces that would represent the evolution of the universe in time. One would still have to say how the possible histories of the universe would behave at the boundary of space–time in the past. We do not and cannot know the boundary conditions of the universe in the past. However, one could avoid this difficulty if the boundary condition of the universe is that it has no boundary. In other words, all the possible histories are finite in extent but have no boundaries, edges, or singularities. They are like the surface of the Earth, but with two more dimensions. In that case, the beginning of time would be a regular smooth point of space–time. This means that the universe would have begun its expansion in a very smooth and ordered state. It could not have been completely uniform because that would violate the uncertainty principle of quantum theory. There had to be small fluctuations in the density and velocities of particles. The no boundary condition, however, would imply that these fluctuations were as small as they could be, consistent with the uncertainty principle.

The universe would have started off with a period of exponential or "inflationary" expansion. In this, it would have increased its size by a very large factor. During this expansion, the density fluctuations would have remained small at first, but later would have started to grow. Regions in which the density was slightly higher than average would have had their expansion slowed down by the gravitational attraction of the extra mass. Eventually, such regions would stop expanding and would collapse to form galaxies, stars, and beings like us.

The universe would have started off with a period of exponential or "inflationary" expansion. In this, it would have increased its size by a very large factor.

The collapse of

a star to form a

black hole is rather

like the later stages

of the collapse of

the whole universe.

The universe would have started in a smooth and ordered state and would become lumpy and disordered as time went on. This would explain the existence of the thermodynamic arrow of time. The universe would start in a state of high order and would become more disordered with time. As I showed earlier, the psychological arrow of time points in the same direction as the thermodynamic arrow. Our subjective sense of time would therefore be that in which the universe is expanding, rather than the opposite direction, in which it would be contracting.

DOES THE ARROW OF TIME REVERSE?

But what would happen if and when the universe stopped expanding and began to contract again? Would the thermodynamic arrow reverse and disorder begin to decrease with time? This would lead to all sorts of science–fiction–like possibilities for people who survived from the expanding to the contracting phase. Would they see broken cups gathering themselves together off the floor and jumping back on the table? Would they be able to remember tomorrow's prices and make a fortune on the stock market?

It might seem a bit academic to worry about what would happen when the universe collapses again, as it will not start to contract for at least another ten thousand million years. But there is a quicker way to find out what will happen: Jump into a black hole. The collapse of a star to form a black hole is rather like the later stages of the collapse of the whole universe. Thus, if disorder were to decrease in the contracting phase of the universe, one might also expect it to decrease inside a black hole. So perhaps an astronaut who fell into a black hole would be able to make money at roulette by remembering where the ball went before he placed his bet. Unfortunately, however, he would not have long to play before he was turned to spaghetti by the very strong gravitational fields. Nor would he be able to let us know about the reversal of the thermodynamic arrow, or even bank his winnings, because he would be trapped behind the event horizon of the black hole.

At first, I believed that disorder would decrease when the universe recollapsed. This was because I thought that the universe had to return to a smooth and ordered state when it became small again. This would have meant that the contracting phase was like the time reverse of the expanding phase. People in the contracting phase would live their lives backward. They would die before they were born and would get younger as the universe contracted. This idea is attractive because it would mean a nice symmetry between the expanding and contracting phases. However, one cannot adopt it on its own, independent of other ideas about the universe. The question is: Is it implied by the no boundary condition or is it inconsistent with that condition?

As I mentioned, I thought at first that the no boundary condition did indeed imply that disorder would decrease in the contracting phase. This was based on work on a simple model of the universe in which the collapsing phase looked like the time reverse of the expanding phase. However, a colleague of mine, Don Page, pointed out that the no boundary condition did not require the contracting phase necessarily to be the time reverse of the expanding phase. Further, one of my students, Raymond Laflamme, found that in a slightly more complicated model, the collapse of the universe was very different from the expansion. I realized that I had made a mistake. In fact, the no boundary condition implied that disorder would continue to increase during the contraction. The thermodynamic and psychological arrows of time would not reverse when the universe begins to recontract, or inside black holes.

The no boundary condition did not require the contracting phase necessarily to be the time reverse of the expanding phase.

What should you do when you find you have made a mistake like that? Some people, like Eddington, never admit that they are wrong. They continue to find new, and often mutually inconsistent, arguments to support their case. Others claim to have never really supported the incorrect view in the first place or, if they did, it was only to show that it was inconsistent. I could give a large number of examples of this, but I won't because it would make me too unpopular. It seems to me much better and less confusing if you admit in print that you were wrong. A good example of this was Einstein, who said that the cosmological constant, which he introduced when he was trying to make a static model of the universe, was the biggest mistake of his life.

The THEORY of EVERYTHING

It would be very difficult to construct a complete unified theory of everything all at one go. So instead we have made progress by finding partial theories. These describe a limited range of happenings and neglect other effects, or approximate them by certain numbers. In chemistry, for example, we can calculate the interactions of atoms without knowing the internal structure of the nucleus of an atom. Ultimately, however, one would hope to find a complete, consistent, unified theory that would include all these partial theories as approximations. The quest for such a theory is known as "the unification of physics."

Einstein spent most of his later years unsuccessfully searching for a unified theory, but the time was not ripe: Very little was known about the nuclear forces. Moreover, Einstein refused to believe in the reality of quantum mechanics, despite the important role he had played in its development. Yet it seems that the uncertainty principle is a fundamental feature of the universe we live in. A successful unified theory must therefore necessarily incorporate this principle.

The prospects for finding such a theory seem to be much better now because we know so much more about the universe. But we must beware of overconfidence. We have had false dawns before. At the beginning of this century, for example, it was thought that everything could be explained in terms of the properties of continuous matter, such as elasticity and heat conduction. The discovery of atomic structure and the uncertainty principle put an end to that.

Then again, in 1928, Max Born told a group of visitors to Göttingen University, "Physics, as we know it, will be over in six

It seems that the uncertainty principle is a fundamental feature of the universe we live in. A successful unified theory must therefore necessarily incorporate this principle.

months." His confidence was based on the recent discovery by Dirac of the equation that governed the electron. It was thought that a similar equation would govern the proton, which was the only other particle known at the time, and that would be the end of theoretical physics. However, the discovery of the neutron and of nuclear forces knocked that one on the head, too.

I still believe there are grounds for cautious optimism that we may now be near the end of the search for the ultimate laws of nature.

Having said this, I still believe there are grounds for cautious optimism that we may now be near the end of the search for the ultimate laws of nature. At the moment, we have a number of partial theories. We have general relativity, the partial theory of gravity, and the partial theories that govern the weak, the strong, and the electromagnetic forces. The last three may be combined in so-called grand unified theories. These are not very satisfactory because they do not include gravity. The main difficulty in finding a theory that unifies gravity with the other forces is that general relativity is a classical theory. That is, it does not incorporate the uncertainty principle of quantum mechanics. On the other hand, the other partial theories depend on quantum mechanics in an essential way. A necessary first step, therefore, is to combine general relativity with the uncertainty principle. As we have seen, this can produce some remarkable consequences, such as black holes not being black, and the universe being completely self–contained and without boundary. The trouble is, the uncertainty principle means that even empty space is filled with pairs of virtual particles and antiparticles. These pairs would have an infinite amount of energy. This means that their gravitational attraction would curve up the universe to an infinitely small size.

Rather similar, seemingly absurd infinities occur in the other quantum theories. However, in these other theories, the infinities can be canceled out by a process called renormalization. This involves adjusting the masses of the particles and the strengths of the forces in the theory by an infinite amount. Although this technique is rather dubious mathematically, it does seem to work in practice. It has been used to make predictions that agree with observations to an extraordinary degree of accuracy.

Renormalization, however, has a serious drawback from the point of view of trying to find a complete theory. When you subtract infinity from infinity, the answer can be anything you want. This means that the actual values of the masses and the strengths of the forces cannot be predicted from the theory. Instead, they have to be chosen to fit the observations. In the case of general relativity, there are only two quantities that can be adjusted: the strength of gravity and the value of the cosmological constant. But adjusting these is not sufficient to remove all the infinities. One therefore has a theory that seems to predict that certain quantities, such as the curvature of space–time, are really infinite, yet these quantities can be observed and measured to be perfectly finite. In an attempt to overcome this problem, a theory called "supergravity" was suggested in 1976. This theory was really just general relativity with some additional particles.

In general relativity, the gravitational force can be thought of as being carried by a particle of spin 2 called the graviton. The idea was to add certain other new particles of spin 3/2, 1, 1/2, and 0. In a sense, all these particles could then be regarded as different aspects of the same "superparticle." The virtual particle/antiparticle pairs of spin 1/2 and 3/2 would have negative energy. This would tend to cancel out the positive energy of the virtual pairs of particles of spin 0, 1, and 2. In this way, many of the possible infinities would cancel out, but it was suspected that some infinities might still remain. However, the calculations required to find out whether there were any infinities left uncanceled were so long and difficult that no one was prepared to undertake them. Even with a computer it was reckoned it would take at least four years. The chances were very high that one would make at least one mistake, and probably more. So one would know one had the right answer only if someone else repeated the calculation and got the same answer, and that did not seem very likely.

Because of this problem, there was a change of opinion in favor of what are called string theories. In these theories the basic objects are not particles that occupy a single point of

In the case of general relativity, there are only two quantities that can be adjusted: the strength of gravity and the value of the cosmological constant.

space. Rather, they are things that have a length but no other dimension, like an infinitely thin loop of string. A particle occupies one point of space at each instant of time. Thus, its history can be represented by a line in space-time called the "world–line." A string, on the other hand, occupies a line in space at each moment of time. So its history in space–time is a two–dimensional surface called the "world–sheet." Any point on such a world–sheet can be described by two numbers, one specifying the time and the other the position of the point on the string. The world-sheet of a string is a cylinder or tube. A slice

(Top) The world-sheet of a string is a cylinder or tube. (Bottom) Two pieces of string can join together to form a single string.

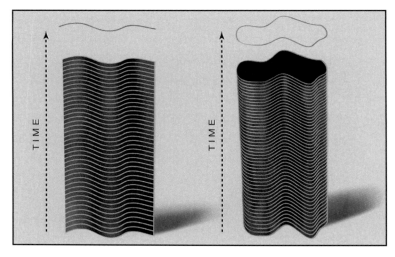

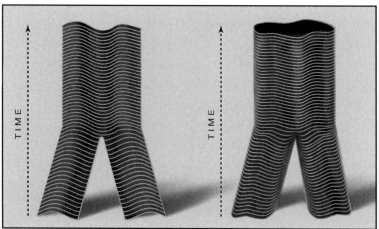

through the tube is a circle, which represents the position of the string at one particular time.

Two pieces of string can join together to form a single string. It is like the two legs joining on a pair of trousers. Similarly, a single piece of string can divide into two strings. In string theories, what were previously thought of as particles are now pictured as waves traveling down the string, like waves on a washing line. The emission or absorption of one particle by another corresponds to the dividing or joining together of strings. For example, the gravitational force of the sun on the Earth corresponds to an H-shaped tube or pipe. String theory is rather like plumbing, in a way. Waves on the two vertical sides of the H correspond to the particles in the sun and the Earth, and waves on the horizontal crossbar correspond to the gravitational force that travels between them.

String theory has a curious history. It was originally invented in the late 1960s in an attempt to find a theory to describe the strong force. The idea was that particles like the proton and the neutron could be regarded as waves on a string. The strong forces between the particles would correspond to pieces of string that went between other bits of string, like in a spider's web. For this theory to give the observed value of the strong force between particles, the strings had to be like rubber bands with a pull of about ten tons.

In 1974 Joël Scherk and John Schwarz published a paper in which they showed that string theory could describe the gravitational force, but only if the tension in the string were very much higher—about 10^{39} tons. The predictions of the string theory would be just the same as those of general relativity on normal length scales, but they would differ at very small distances—less than 10^{-33} centimeters. Their work did not receive much attention, however, because at just about that time, most people abandoned the original string theory of the strong force. Scherk died in tragic circumstances. He suffered from diabetes and went into a coma when no one was around to give him an

In string theories, what were previously thought of as particles are now pictured as waves traveling down the string, like waves on a washing line.

The science fiction idea is that one can take a shortcut through a higher dimension.

injection of insulin. So Schwarz was left alone as almost the only supporter of string theory, but now with a much higher proposed value of the string tension.

There seemed to have been two reasons for the sudden revival of interest in strings in 1984. One was that people were not really making much progress toward showing that supergravity was finite or that it could explain the kinds of particles that we observe. The other was the publication of a paper by John Schwarz and Mike Green which showed that string theory might be able to explain the existence of particles that have a built–in left–handedness, like some of the particles that we observe. Whatever the reasons, a large number of people soon began to work on string theory. A new version was developed, the so–called heterotic string. This seemed as if it might be able to explain the types of particle that we observe.

String theories also lead to infinities, but it is thought they will all cancel out in versions like the heterotic string.

Imagine that the space we lived in had only two dimensions and was curved like the surface of a doughnut. If you were able to travel in the third dimension, you could cut straight across instead of going around the doughnut to get to the other side.

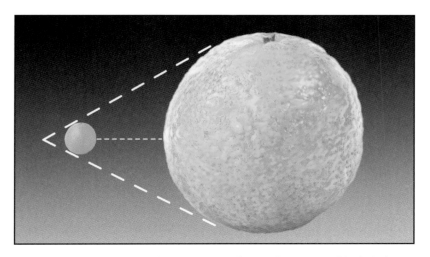

We don't see the extra dimensions in space because they are so small we don't notice them. Similarly, we don't see the bumps and wrinkles of an orange when we look at it from a distance.

String theories, however, have a bigger problem. They seem to be consistent only if space–time has either ten or twenty–six dimensions, instead of the usual four. Of course, extra space–time dimensions are a commonplace of science fiction; indeed, they are almost a necessity. Otherwise, the fact that relativity implies that one cannot travel faster than light means that it would take far too long to get across our own galaxy, let alone to travel to other galaxies. The science fiction idea is that one can take a shortcut through a higher dimension. One can picture this in the following way. Imagine that the space we live in had only two dimensions and was curved like the surface of a doughnut or a torus. If you were on one side of the ring and you wanted to get to a point on the other side, you would have to go around the ring. However, if you were able to travel in the third dimension, you could cut straight across.

Why don't we notice all these extra dimensions if they are really there? Why do we see only three space and one time dimension? The suggestion is that the other dimensions are curved up into a space of very small size, something like a million million million million millionth of an inch. This is so small that we just don't notice it. We see only the three space and one time dimension in which space-time is thoroughly flat. It is like the surface of an orange: if you look at it close up, it is all curved and wrinkled, but if you look at it from a distance, you don't see the bumps, and it appears to be smooth. So it is with space–time. On a very small scale, it is ten–dimensional and highly curved. But on bigger scales, you don't see the curvature or the extra dimensions.

If this picture is correct, it spells bad news for would-be space travelers. The extra dimensions would be far too small to allow a spaceship through. However, it raises another major problem. Why should some, but not all, of the dimensions be curled up into a small ball? Presumably, in the very early universe, all the dimensions would have been very curved. Why did three space and one time dimension flatten out, while the other dimensions remained tightly curled up?

One possible answer is the anthropic principle. Two space dimensions do not seem to be enough to allow for the development of complicated beings like us. For example, two–dimensional people living on a one-dimensional Earth would have to climb over each other in order to get past each other. If a two-dimensional creature ate something it could not digest completely, it would have to bring up the remains the same way it swallowed them, because if there were a passage through its body, it would divide the creature into two separate parts. Our two–dimensional being would fall apart. Similarly, it is difficult to see how there could be any circulation of the blood in a two-dimensional creature. There would also be problems with more than three space dimensions. The gravitational force between two bodies would decrease more rapidly with distance than it does in three dimensions. The significance of this is that the orbits of planets, like the Earth, around the sun would be unstable. The least disturbance from a circular orbit, such as would be caused by the gravitational attraction of other planets, would cause the Earth to spiral away from or into the sun. We would either freeze or be burned up. In fact, the same behavior of gravity with distance would mean that the sun would also be unstable. It

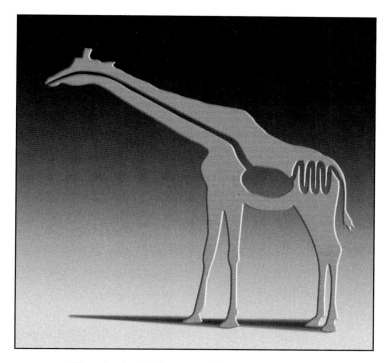

The anthropic principle points out that two space dimensions are not enough for complicated beings like humans and giraffes.

would either fall apart or it would collapse to form a black hole. In either case, it would not be much use as a source of heat and light for life on Earth. On a smaller scale, the electrical forces that cause the electrons to orbit around the nucleus in an atom would behave in the same way as the gravitational forces. Thus, the electrons would either escape from the atom altogether or it would spiral into the nucleus. In either case, one could not have atoms as we know them.

It seems clear that life, at least as we know it, can exist only in regions of space-time in which three space and one time dimension are not curled up small. This would mean that one could appeal to the anthropic principle, provided one could show that string theory does at least allow there to be such regions of the universe. And it seems that indeed each string theory does allow such regions. There may well be other regions of the universe, or other universes (whatever that may mean) in which all the dimensions are curled up small, or in which more than

It is likely that answers to these questions will be found over the next few years, and that by the end of the century we shall know whether string theory is indeed the long sought-after unified theory of physics.

four dimensions are nearly flat. But there would be no intelligent beings in such regions to observe the different number of effective dimensions.

Apart from the question of the number of dimensions that space-time appears to have, string theory still has several other problems that must be solved before it can be acclaimed as the ultimate unified theory of physics. We do not yet know whether all the infinities cancel each other out, or exactly how to relate the waves on the string to the particular types of particle that we observe. Nevertheless, it is likely that answers to these questions will be found over the next few years, and that by the end of the century we shall know whether string theory is indeed the long sought-after unified theory of physics.

Can there really be a unified theory of everything? Or are we just chasing a mirage? There seem to be three possibilities:

- There really is a complete unified theory, which we will someday discover if we are smart enough.
- There is no ultimate theory of the universe, just an infinite sequence of theories that describe the universe more and more accurately.
- There is no theory of the universe. Events cannot be predicted beyond a certain extent but occur in a random and arbitrary manner.

Some would argue for the third possibility on the grounds that if there were a complete set of laws, that would infringe on God's freedom to change His mind and to intervene in the world. It's a bit like the old paradox: Can God make a stone so heavy that He can't lift it? But the idea that God might want to change His mind is an example of the fallacy, pointed out by St. Augustine, of imagining God as a being existing in time. Time is a property only of the universe that God created. Presumably, He knew what He intended when He set it up.

With the advent of quantum mechanics, we have come to realize that events cannot be predicted with complete accuracy but that there is always a degree of uncertainty. If one liked, one could ascribe this randomness to the intervention of God. But it would be a very strange kind of intervention. There is no evidence that it is directed toward any purpose. Indeed, if it were, it wouldn't be random. In modern times, we have effectively removed the third possibility by redefining the goal of science. Our aim is to formulate a set of laws that will enable us to predict events up to the limit set by the uncertainty principle.

The second possibility, that there is an infinite sequence of more and more refined theories, is in agreement with all our experience so far. On many occasions, we have increased the sensitivity of our measurements or made a new class of observations only to discover new phenomena that were not predicted by the existing theory. To account for these, we have had to develop a more advanced theory. It would therefore not be very surprising if we find that our present grand unified theories break down when we test them on bigger and more powerful particle accelerators. Indeed, if we didn't expect them to break down, there wouldn't be much point in spending all that money on building more powerful machines.

However, it seems that gravity may provide a limit to this sequence of "boxes within boxes." If one had a particle with an energy above what is called the Planck energy, 10^{19} GeV, its mass would be so concentrated that it would cut itself off from the rest of the universe and form a little black hole. Thus, it does seem that the sequence of more and more refined theories should have some limit as we go to higher and higher energies. There should be some ultimate theory of the universe. Of course, the Planck energy is a very long way from the energies of around a GeV, which are the most that we can produce in the laboratory at the present time. To bridge that gap would require a particle accelerator that was bigger than the solar system. Such an accelerator would be unlikely to be funded in the present economic climate.

The second possibility, that there is an infinite sequence of more and more refined theories, is in agreement with all our experience so far.

In Newton's time

it was possible for

an educated person

to have a grasp of

the whole of human

knowledge, at least

in outline. But ever

since then, the pace

of development of

science has made

this impossible.

However, the very early stages of the universe are an arena where such energies must have occurred. I think that there is a good chance that the study of the early universe and the requirements of mathematical consistency will lead us to a complete unified theory by the end of the century—always presuming we don't blow ourselves up first.

What would it mean if we actually did discover the ultimate theory of the universe? It would bring to an end a long and glorious chapter in the history of our struggle to understand the universe. But it would also revolutionize the ordinary person's understanding of the laws that govern the universe. In Newton's time it was possible for an educated person to have a grasp of the whole of human knowledge, at least in outline. But ever since then, the pace of development of science has made this impossible. Theories were always being changed to account for new observations. They were never properly digested or simplified so that ordinary people could understand them. You had to be a specialist, and even then you could only hope to have a proper grasp of a small proportion of the scientific theories.

Further, the rate of progress was so rapid that what one learned at school or university was always a bit out of date. Only a few people could keep up with the rapidly advancing frontier of knowledge. And they had to devote their whole time to it and specialize in a small area. The rest of the population had little idea of the advances that were being made or the excitement they were generating.

Seventy years ago, if Eddington is to be believed, only two people understood the general theory of relativity. Nowadays tens of thousands of university graduates understand it, and many millions of people are at least familiar with the idea. If a complete unified theory were discovered, it would be only a matter of time before it was digested and simplified in the same way. It could then be taught in schools, at least in outline. We would then all be able to have some understanding of the laws that govern the universe and which are responsible for our existence.

Einstein once asked a question: "How much choice did God have in constructing the universe?" If the no boundary proposal is correct, He had no freedom at all to choose initial conditions. He would, of course, still have had the freedom to choose the laws that the universe obeyed. This, however, may not really have been all that much of a choice. There may well be only one or a small number of complete unified theories that are self-consistent and which allow the existence of intelligent beings.

We can ask about the nature of God even if there is only one possible unified theory that is just a set of rules and equations. What is it that breathes fire into the equations and makes a universe for them to describe? The usual approach of science of constructing a mathematical model cannot answer the question of why there should be a universe for the model to describe. Why does the universe go to all the bother of existing? Is the unified theory so compelling that it brings about its own existence? Or does it need a creator, and, if so, does He have any effect on the universe other than being responsible for its existence? And who created Him?

Up until now, most scientists have been too occupied with the development of new theories that describe what the universe is, to ask the question why. On the other hand, the people whose business it is to ask why—the philosophers—have not been able to keep up with the advance of scientific theories. In the eighteenth century, philosophers considered the whole of human knowledge, including science, to be their field. They discussed questions such as: Did the universe have a beginning? However, in the nineteenth and twentieth centuries, science became too technical and mathematical for the philosophers or anyone else, except a few specialists. Philosophers reduced the scope of their inquiries so much that Wittgenstein, the most famous philosopher of this century, said, "The sole remaining task for philosophy is the analysis of language." What a comedown from the great tradition of philosophy from Aristotle to Kant.

We can ask about the nature of God even if there is only one possible unified theory that is just a set of rules and equations.

However, if we do discover a complete theory, it should in time be understandable in broad principle by everyone, not just a few scientists. Then we shall all be able to take part in the discussion of why the universe exists. If we find the answer to that, it would be the ultimate triumph of human reason. For then we would know the mind of God.

INDEX

CREDITS

Illustration Credits

Illustration on pages 7, 28, 73, 78, 82, 88, 100, 102, 103, and 105 conceived and
    created by Rob Fiore.

Photo Credits

Page 2: Historical Cosmologies-Sheila Terry / Science photo Library

Page 6: Quintuplet Cluster (1999)-NASA, Don Figer, STScI

Page 12: Fireworks of Star Formation Light Up a Galaxy-NASA, The Hubble Heritage Team.

Page 13: Faint Blue Galaxy (1995) - Rogier Windhorst and Simon Driver.
    (Arizona State University), Bill Keel (University of Alabama), and NASA

Page 16: Rocket-NASA

Page 18: Milky Way (2001) – NASA/Umass/D. Wang et al

Page 29: Einstein Ring-Jon Lomberg/Science Photo Library

Page 30: Young stars forming-NASA

Page 31: Life Cycle of Stars/Supernova (1999)-NASA, Wolfgang Brandner JPL-IPAC,
    Eva K. Grebel University of Washington.

Page 32: Dying Stars (2002) –NASA and the Hubble Heritage Team (STScI/AURA)

Page 35: White Dwarf Stars (2001) – NASA and H. Richer (University of British Columbia),
    credit for ground-based photo NOAO/AURA/NSF.

Page 36: Gravity and Stars (1999) – NASA, The Hubble Heritage Team, STScI/AURA

Page 38: Ways to Grow A Black Hole-K. Cordes & S. Brown (STScI).

Page 41: Astronaut (1994) – NASA.

Page 43: Black Hole Mass (2000) – NASA and Karl Gebhardt (Lick Observatory).

Page 53: Event Horizon (2001) – Greg Bacon (STScI/AVL)

Page 59: Black Hole Emission (1990) – Dana Berry (STScI)

Page 62: Astronaut II (1994) – NASA

Page 68: Spiral Galaxy – NASA, The Hubble Heritage Team, STScI, AURA

Page 80: Earth Surface (2002) – NASA

Page 84: Fireworks Finale of Stars (2002) – NASA and K. Lanzetta (SUNY), artwork by
    A. Schaller for STScI.